GRAPHING
CALCULATOR MANUAL

JUDITH A. PENNA

Indiana University Purdue University Indianapolis

CALCULUS
AND ITS APPLICATIONS

Marvin L. Bittinger

Indiana University Purdue University Indianapolis

PEARSON

Addison
Wesley

Boston   San Francisco   New York
London   Toronto   Sydney   Tokyo   Singapore   Madrid
Mexico City   Munich   Paris   Cape Town   Hong Kong   Montreal

ISBN   0-321-17312-0

1 2 3 4 5 6 VHG 06 05 04 03

Contents

The TI-83 and TI-83 Plus
Graphics Calculators

Chapter 1
Functions, Graphs, and Models

GETTING STARTED

Press $\boxed{\text{ON}}$ to turn on the TI-83 or TI-83+ graphing calculator. ($\boxed{\text{ON}}$ is the key at the bottom left-hand corner of the keypad.) You should see a blinking rectangle, or cursor, on the screen. If you do not see the cursor, try adjusting the display contrast. To do this, first press $\boxed{\text{2nd}}$. ($\boxed{\text{2nd}}$ is the yellow key in the left column of the keypad.) Then press and hold $\boxed{\triangle}$ to increase the contrast or $\boxed{\triangledown}$ to decrease the contrast. If the contrast needs to be adjusted further after the first adjustment, press $\boxed{\text{2nd}}$ again then then hold $\boxed{\triangle}$ or $\boxed{\triangledown}$ to increase or decrease the contrast, respectively.

To turn the grapher off, press $\boxed{\text{2nd}}$ $\boxed{\text{OFF}}$. (OFF is the second operation associated with the $\boxed{\text{ON}}$ key.) The grapher will turn itself off automatically after about five minutes without any activity.

Press $\boxed{\text{MODE}}$ to display the MODE settings. Initially you should select the settings on the left side of the display.

To change a setting on the Mode screen use $\boxed{\triangledown}$ or $\boxed{\triangle}$ to move the cursor to the line of that setting. Then use $\boxed{\triangleright}$ or $\boxed{\triangleleft}$ to move the blinking cursor to the desired setting and press $\boxed{\text{ENTER}}$. Press $\boxed{\text{CLEAR}}$ or $\boxed{\text{2nd}}$ $\boxed{\text{QUIT}}$ to leave the MODE screen. (QUIT is the second operation associated with the $\boxed{\text{MODE}}$ key.) In general, second operations are written in yellow above the keys on the keypad. Pressing $\boxed{\text{CLEAR}}$ or $\boxed{\text{2nd}}$ $\boxed{\text{QUIT}}$ will take you to the home screen where computations are performed.

The TI-83 and TI-83 Plus graphing calculators are very similar in many respects. For that reason, most of the keystrokes and instructions presented in this section of the graphing calculator manual will apply to both graphers. Where they differ, keystrokes and instructions for using the TI-83 will be given first, followed by those for the TI-83 Plus.

It will be helpful to read the Getting Started section of the Texas Instruments Guidebook that was packaged with your graphing calculator before proceeding.

SETTING THE VIEWING WINDOW

Section 1.1, page 7 (Page numbers refer to pages in the textbook.)

The viewing window is the portion of the coordinate plane that appears on the grapher's screen. It is defined by the minimum and maximum values of x and y: Xmin, Xmax, Ymin, and Ymax. The notation [Xmin, Xmax, Ymin, Ymax] is used in the text to represent these window settings or dimensions. For example, $[-12, 12, -8, 8]$ denotes a window that displays the portion of the x-axis from -12 to 12 and the portion of the y-axis from -8 to 8. In addition, the distance between tick marks on the axes is defined by the settings Xscl and Yscl. In this manual Xscl and Yscl will be assumed to be 1 unless noted otherwise. The setting Xres sets the pixel resolution. We usually select Xres = 1. The window corresponding to the settings $[-20, 30, -12, 20]$, Xscl = 5, Yscl = 2, Xres = 1, is shown below.

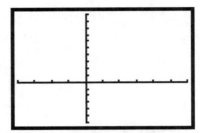

Press the WINDOW key on the top row of the keypad to display the current window settings on your grapher. The standard settings $[-10, 10, -10, 10]$, Xscl = 1, Yscl = 1, are shown below.

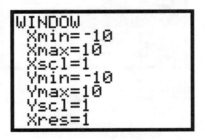

To change a setting, position the cursor beside the setting you wish to change and enter the new value. For example, to change from the standard settings to $[-20, 30, -12, 20]$, Xscl = 5, Yscl = 2, on the WINDOW screen press (−) 2 0 ENTER 3 0 ENTER 5 ENTER (−) 1 2 ENTER 2 0 ENTER 2 ENTER. You must use the (−) key on the bottom of the keypad rather than the − key in the right-hand column to enter a negative number. (−) represents "the opposite of" or "the additive inverse of" whereas − is the key for the subtraction operation. The ▽ key may be used instead of ENTER after typing each window setting. To see the window shown above, press the GRAPH key on the top row of the keypad.

QUICK TIP: To return quickly to the standard window setting $[-10, 10, -10, 10]$, Xscl = 1, Yscl = 1, press ZOOM 6.

GRAPHS

After entering an equation and setting a viewing window, you can view the graph of an equation.

Section 1.1, page 8 Graph $y = x^3 - 5x + 1$ using a graphing calculator.

Equations are entered on the Y =, or equation-editor, screen. Press $\boxed{\text{Y} =}$ to access this screen. If any of Plot 1, Plot 2, and Plot 3 is turned on (highlighted), turn it off by using the arrow keys to move the blinking cursor over the plot name and pressing $\boxed{\text{ENTER}}$. If there is currently an expression displayed for Y_1, clear it by positioning the cursor beside "$Y_1 =$" and pressing $\boxed{\text{CLEAR}}$. Do the same for expressions that appear on all other lines by using $\boxed{\triangledown}$ to move to a line and then pressing $\boxed{\text{CLEAR}}$. Then use $\boxed{\triangle}$ or $\boxed{\triangledown}$ to move the cursor beside "$Y_1 =$." Now press $\boxed{\text{X, T, }\Theta, n}$ $\boxed{\wedge}$ 3 $\boxed{-}$ 5 $\boxed{\text{X, T, }\Theta, n}$ $\boxed{+}$ 1 to enter the right-hand side of the equation in the Y = screen.

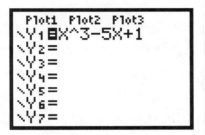

The standard $[-10, 10, -10, 10]$ window is a good choice for this graph. Either enter these dimensions in the WINDOW screen and then press $\boxed{\text{GRAPH}}$ to see the graph or simply press $\boxed{\text{ZOOM}}$ 6 to select the standard window and see the graph.

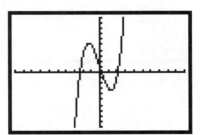

You can edit your entry if necessary. If, for instance, you pressed 6 instead of 5, use the $\boxed{\triangleleft}$ key to move the cursor to 6 and then press 5 to overwrite it. If you forgot to type the plus sign, move the cursor to 1, then press $\boxed{\text{2nd}}$ $\boxed{\text{INS}}$ $\boxed{+}$ to insert the plus sign before the 1. (INS is the second operation associated with the $\boxed{\text{DEL}}$ key.) You can continue to insert symbols immediately after the first insertion without pressing $\boxed{\text{2nd}}$ $\boxed{\text{INS}}$ again. If you typed 52 instead of 5, move the cursor to 2 and press $\boxed{\text{DEL}}$ to delete the 2.

An equation must be solved for y before it can be graphed on the TI-83 or the TI-83 Plus.

Section 1.1, page 8 To graph $3x + 5y = 10$, first solve for y, obtaining $y = \dfrac{-3x + 10}{5}$. Then press $\boxed{\text{Y} =}$ and clear any expressions that currently appear. Position the cursor beside "$Y_1 =$." Now press $\boxed{(}$ $\boxed{(-)}$ 3 $\boxed{\text{X, T, }\Theta, n}$ $\boxed{+}$ 1 0 $\boxed{)}$ $\boxed{\div}$ 5 to enter the right-hand side of the equation. Note that without the parentheses the expression $-3x + \dfrac{10}{5}$ would have been entered.

Select a viewing window and then press $\boxed{\text{GRAPH}}$ to display the graph. You may change the viewing window as desired to reveal more or less of the graph. The standard window is shown here.

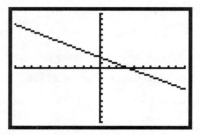

To graph $x = y^2$, first solve the equation for y: $y = \pm\sqrt{x}$. To obtain the entire graph of $x = y^2$, you must graph $y_1 = \sqrt{x}$ and $y_2 = -\sqrt{x}$ on the same screen. Press $\boxed{\text{Y} =}$ and clear any expressions that currently appear. With the cursor beside "$Y_1 =$" press $\boxed{\text{2nd}}$ $\boxed{\sqrt{}}$ $\boxed{\text{X, T, } \Theta, n}$ $\boxed{)}$. ($\boxed{\sqrt{}}$ is the second operation associated with the $\boxed{x^2}$ key.) A left parenthesis appears along with the radical symbol, so a separate keystroke is not necessary to introduce it.

Now use $\boxed{\triangledown}$ to move the cursor beside "Y2 =." We will show two ways to enter $y_2 = -\sqrt{x}$. One is to enter the expression $-\sqrt{x}$ directly by pressing $\boxed{(-)}$ $\boxed{\text{2nd}}$ $\boxed{\sqrt{}}$ $\boxed{\text{X, T, } \Theta, n}$ $\boxed{)}$.

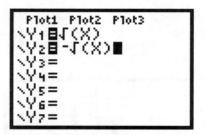

The other method of entering y_2 is based on the observation that $-\sqrt{x}$ is the opposite of the expression for y_1. That is, $y_2 = -y_1$. To enter this press $\boxed{(-)}$ $\boxed{\text{VARS}}$ $\boxed{\triangleright}$ to select Y-Vars. Then press 1 1 to select y_1.

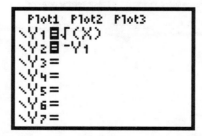

Select a viewing window and press $\boxed{\text{GRAPH}}$ to display the graph. The window shown here is $[-2, 10, -5, 5]$.

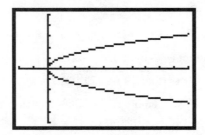

The top half is the graph of y_1, the bottom half is the graph of y_2, and together they yield the graph of $x = y^2$.

THE TABLE FEATURE

For an equation entered in the equation-editor screen, a table of x-and y-values can be displayed.

Section 1.2, page 18 Create a table of ordered pairs for the function $f(x) = x^3 - 5x + 1$.

Enter the function as $y_1 = x^3 - 5x + 1$ as described on page 5 of this manual. Once the equation in entered, press $\boxed{\text{2nd}}$ $\boxed{\text{TBLSET}}$ to display the table set-up screen. (TBLSET is the second function associated with the $\boxed{\text{WINDOW}}$ key.) A minimum value of x can be chosen along with an increment for the x-value. To select a minimum x-value of 0.3 and an increment of 1, press $\boxed{\;.\;}$ 3 $\boxed{\triangledown}$ 1. The "Indpnt" and "Depend" settings should both be "Auto." If either is not, use the $\boxed{\triangledown}$ key to position the blinking cursor over "Auto" on that line and then press $\boxed{\text{ENTER}}$. To display the table press $\boxed{\text{2nd}}$ $\boxed{\text{TABLE}}$. (TABLE is the second operation associated with the $\boxed{\text{GRAPH}}$ key.)

Use the $\boxed{\triangledown}$ and $\boxed{\triangle}$ keys to scroll through the table. For example, by using $\boxed{\triangledown}$ to scroll down we can see that $y_1 = 758.86$ when $x = 9.3$. Using $\boxed{\triangle}$ to scroll up, observe that $y_1 = -31.15$ when $x = -3.7$.

GRAPHS AND FUNCTION VALUES

Section 1.2, page 21 There are several ways to evaluate a function using a grapher. Three of them are described here. Given the function $f(x) = 2x^2 + x$, we will find $f(-2)$. First press $\boxed{Y =}$ and enter the function as $y_1 = 2x^2 + x$. Now we will find $f(-2)$ using the TABLE feature. Press $\boxed{\text{2nd}}$ $\boxed{\text{TBLSET}}$ and select ASK mode by moving the cursor to "Indpnt: Ask" and pressing $\boxed{\text{ENTER}}$. In ASK mode, you supply the x-values and the grapher returns the corresponding y-values. The settings for TblStart nd ΔTbl are irrelevant in this mode. Press $\boxed{\text{2nd}}$ $\boxed{\text{TABLE}}$ and find $f(-2)$ by pressing $\boxed{(-)}$ 2 $\boxed{\text{ENTER}}$. We see that $y_1 = 6$ when $x = -2$, so $f(-2) = 6$.

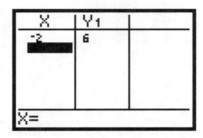

We can also use the VALUE feature from the CALC menu to find $f(-2)$. To do this, graph $y_1 = 2x^2 + x$ in a window that includes the x-value -2. We will use the standard window. Then press $\boxed{\text{2nd}}$ $\boxed{\text{CALC}}$ 1 or $\boxed{\text{2nd}}$ $\boxed{\text{CALC}}$ $\boxed{\text{ENTER}}$ to select the VALUE feature. (CALC is the second operation associated with the $\boxed{\text{TRACE}}$ key in the top row of the keypad.) Now supply the desired x-value by pressing $\boxed{(-)}$ 2. Press $\boxed{\text{ENTER}}$ to see X = -2, Y = 6 at the bottom of the screen. Thus $f(-2) = 6$.

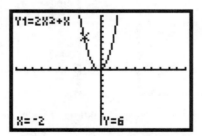

A third method for finding $f(-2)$ uses function notation directly. With $y_1 = 2x^2 + x$ entered on the Y = screen, go to the home screen by pressing $\boxed{\text{2nd}}$ $\boxed{\text{QUIT}}$. (QUIT is the second operation associated with the $\boxed{\text{MODE}}$ key.) Now enter $Y_1(-2)$ by pressing $\boxed{\text{VARS}}$ $\boxed{\triangleright}$ 1 1 $\boxed{(}$ $\boxed{(-)}$ 2 $\boxed{)}$ $\boxed{\text{ENTER}}$. Again we see that $y_1(-2) = 6$, or $f(-2) = 6$.

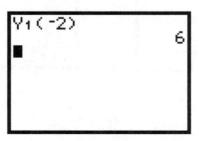

GRAPHING FUNCTIONS DEFINED PIECEWISE; DOT MODE

Section 1.2, Example 9, page 22   Graph: $f(x) = \begin{cases} 4 \text{ for } x \leq 0, \\ 4 - x^2 \text{ for } 0 < x \leq 2, \\ 2x - 6 \text{ for } x > 2. \end{cases}$

We will enter the function using inequality symbols from the TEST menu. Press $\boxed{Y =}$ to go to the equation-editor screen. Clear any entries that are present. Then position the cursor beside $Y_1 =$ and press $\boxed{(}$ $\boxed{4}$ $\boxed{)}$ $\boxed{(}$ $\boxed{X, T, \Theta, n}$ $\boxed{2nd}$ $\boxed{TEST}$ $\boxed{6}$ $\boxed{0}$ $\boxed{)}$ $\boxed{+}$ $\boxed{(}$ $\boxed{4}$ $\boxed{-}$ $\boxed{X, T, \Theta, n}$ $\boxed{x^2}$ $\boxed{)}$ $\boxed{(}$ $\boxed{0}$ $\boxed{2nd}$ $\boxed{TEST}$ $\boxed{5}$ $\boxed{X, T, \Theta, n}$ $\boxed{)}$ $\boxed{(}$ $\boxed{X, T, \Theta, n}$ $\boxed{2nd}$ $\boxed{TEST}$ $\boxed{6}$ $\boxed{2}$ $\boxed{)}$ $\boxed{+}$ $\boxed{(}$ $\boxed{2}$ $\boxed{X, T, \Theta, n}$ $\boxed{-}$ $\boxed{6}$ $\boxed{)}$ $\boxed{(}$ $\boxed{X, T, \Theta, n}$ $\boxed{2nd}$ $\boxed{TEST}$ $\boxed{3}$ $\boxed{2}$ $\boxed{)}$. (TEST is the second operation associated with the $\boxed{MATH}$ key.) The keystrokes $\boxed{2nd}$ $\boxed{TEST}$ 6 open the TEST menu and select item 6, the $\leq$ symbol, from that menu. Similarly, $\boxed{2nd}$ $\boxed{TEST}$ 5 and $\boxed{2nd}$ $\boxed{TEST}$ 3 open the TEST menu and select the symbols $<$ and $>$, respectively.

Now select DOT mode. If this is not done, a vertical line that is not part of the graph will appear. DOT mode can be selected in two ways. One is to press $\boxed{MODE}$, move the cursor over Dot, and press $\boxed{ENTER}$. DOT mode can also be selected on the Y = screen by positioning the cursor over the GraphStyle icon to the left of Y_1 and pressing $\boxed{ENTER}$ repeatedly until the dotted GraphStyle icon appears.

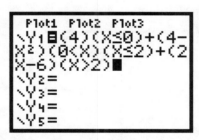

Now choose and enter window dimensions and then press $\boxed{GRAPH}$ to see the graph of the function. It is shown here in the window $[-5, 5, -3, 6]$.

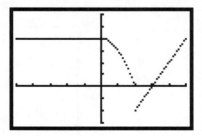

THE TRACE FEATURE

The TRACE feature can be used to display the coordinates of points on a graph.

Section 1.2, page 25 If you selected DOT mode for Example 9 above, return to CONNECTED mode now. Then enter the function $f(x) = x^3 - 5x + 1$ (see page 5 of this manual) and graph it in the window $[-5, 5, -10, 10]$. Now press $\boxed{TRACE}$. A blinking cursor appears on the graph and the coordinates of the point at which it is positioned are displayed at the bottom of the screen. Use the $\boxed{\triangleleft}$ and $\boxed{\triangleright}$ keys to move the cursor along the graph to see the coordinates of other points.

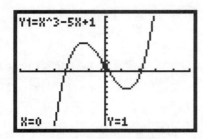   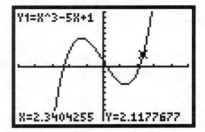

SQUARING THE VIEWING WINDOW

Section 1.4, page 41 In the standard window, the distance between tick marks on the y-axis is about 2/3 the distance between tick marks on the x-axis. It is often desirable to choose window dimensions for which these distances are the same, creating a "square" window. Any window in which the ratio of the length of the y-axis to the length of the x-axis is 2/3 will produce this effect. This can be accomplished by selecting dimensions for which $\text{Ymax} - \text{Ymin} = \dfrac{2}{3}(\text{Xmax} - \text{Xmin})$.

The standard window is shown on the left below and the square window $[-6, 6, -4, 4]$ is shown on the right. Observe that the distance between tick marks appears to be the same on both axes in the square window.

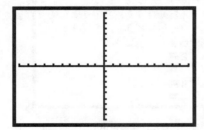   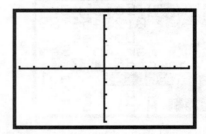

The window can also be squared by pressing $\boxed{\text{ZOOM}}$ 5 to select the ZSquare feature from the ZOOM menu. Starting with the standard window and pressing $\boxed{\text{ZOOM}}$ 5 produces the dimensions and the window shown below.

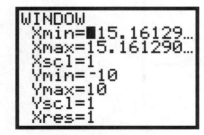   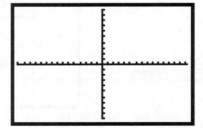

THE INTERSECT FEATURE

We can use the Intersect feature from the CALC menu to solve equations.

Section 1.5, page 61 Solve the equation $x^3 = 3x + 1$ using the Intersect feature.

On the equation editor screen, clear any existing entries and then enter $Y_1 = x^3$ and $Y_2 = 3x + 1$. Now graph these equations in an appropriate window. One good choice is $[-3, 3, -10, 10]$. The solutions of the equation $x^3 = 3x + 1$ are the first coordinates of the points of intersection of these graphs. We will use the Intersect feature to find the leftmost point of intersection first. Press [2nd] [CALC] 5 to select Intersect from the CALC menu. The query "First curve?" appears at the bottom of the screen. The blinking cursor is positioned on the graph of Y_1. This is indicated by the notation $Y_1 = X \wedge 3$ in the upper left-hand corner of the screen. Press [ENTER] to indicate that this is the first curve involved in the intersection. Next the query "Second curve?" appears at the bottom of the screen. The blinking cursor is now positioned on the graph of Y_2 and the notation $Y_2 = 3x + 1$ should appear in the top left-hand corner of the screen. Press [ENTER] to indicate that this is the second curve. We identify the curves for the grapher since we could have as many as ten graphs on the screen at once. After we identify the second curve, the query "Guess?" appears at the bottom of the screen. Use the right and left arrow keys to move the blinking cursor close to the point of intersection of the graphs. This tells the grapher which point of intersection we are trying to find. When the cursor is positioned, press [ENTER] a third time. Now the coordinates of the point of intersection appear at the bottom of the screen.

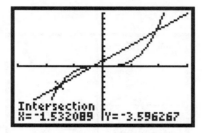

We see that, at the leftmost point of intersection, $x \approx -1.53$, so one solution of the equation is approximately -1.53. Repeat this process two times to find the coordinates of the other two points of intersection. We find that the other two solutions of the equation are approximately -0.35 and 1.88.

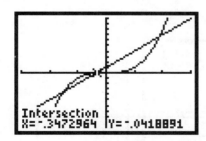   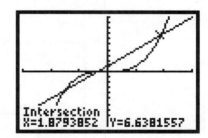

THE ZERO FEATURE

When an equation is expressed in the form $f(x) = 0$, it can be solved using the Zero feature from the CALC menu.

Section 1.5, page 61 Solve the equation $x^3 = 3x + 1$ using the Zero feature.

First subtract $3x$ and 1 on both sides of the equation to obtain an equivalent equation with 0 on one side. We have $x^3 - 3x - 1 = 0$. The solutions of the equation $x^3 = 3x + 1$ are the values of x for which the function $f(x) = x^3 - 3x - 1$ is equal to 0. We can use the Zero feature to find these values, or zeros.

On the equation-editor screen, clear any existing entries and then enter $y_1 = x^3 - 3x - 1$. Now graph the function in a viewing window that shows the x-intercepts clearly. One good choice is $[-3, 3, -5, 8]$. We see that the function has three zeros. They appear to be about -1.5, -0.5, and 2.

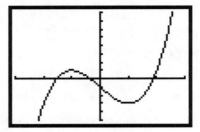

We will find the zero near -1.5 first. Press $\boxed{\text{2nd}}$ $\boxed{\text{CALC}}$ 2 to select the Zero feature from the CALC menu. We are prompted to select a left bound. This means that we must choose an x-value that is to the left of -1.5 on the x-axis. This can be done by using the left- and right-arrow keys to move the cursor to a point on the curve to the left of -1.5 or by keying in a value less than -1.5.

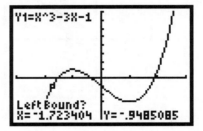

Once this is done press $\boxed{\text{ENTER}}$. Now we are prompted to select a right bound that is to the right of -1.5 on the x-axis. Again, this can be done by using the arrow keys to move the cursor to a point on the curve to the right of -1.5 or by keying in a value greater that -1.5.

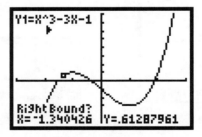

Press $\boxed{\text{ENTER}}$ again. Finally we are prompted to make a guess as to the value of the zero. Move the cursor to a point close to the zero or key in a value.

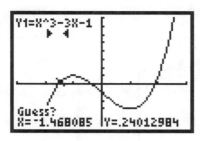

Press ENTER a third time. We see that $y = 0$ when $x \approx -1.53$, so -1.53 is a zero of the function.

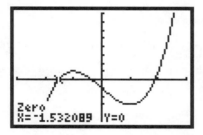

Select Zero from the CALC menu a second time to find the zero near -0.5 and a third time to find the zero near 2. We see that the other two zeros are approximately -0.35 and 1.88.

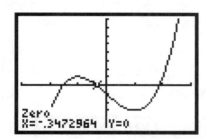 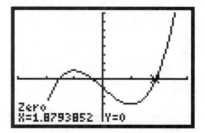

ABSOLUTE-VALUE FUNCTIONS

We can use the absolute-value option to perform computations involving absolute value and to graph absolute-value functions.

Section 1.5, page 65 Graph $f(x) = |x|$.

The absolute-value option can be accessed from either the MATH NUM (Number) menu or from the catalog. Before either option is chosen, first press Y = to go to the equation-editor screen and then clear any existing entries. Now position the cursor beside $Y1 =$ and enter $|x|$ as abs(x).

To do this using the MATH NUM menu, first press MATH ▷ ENTER or MATH ▷ 1 to copy "abs(" to the equation-editor screen. Then press X, T, Θ, n). To select "abs(" from the catalog, press 2nd CATALOG ENTER or 2nd CATALOG 1. (CATALOG is the second operation associated with the 0 numeric key.) Then press X, T, Θ, n). In either case, choose an appropriate viewing window and press GRAPH to see the graph of the function. The graph is shown here in the standard viewing window.

   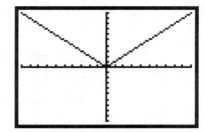

GRAPHING RADICAL FUNCTIONS

There are various way to enter radical expressions on the TI-83 and TI-83 Plus.

Section 1.5, page 67 We discussed entering an expression containing a square root on page 6 of this manual. We can use the $\sqrt[3]{}$ option from the MATH menu to enter an expression containing a cube root. To enter $y_1 = \sqrt[3]{x - 2}$, for example, position the cursor beside Y1 = on the equation-editor screen and press $\boxed{\text{MATH}}$ 4 to select $\sqrt[3]{}$. Then press $\boxed{\text{X, T, }\Theta, n}$ $\boxed{-}$ $\boxed{2}$ $\boxed{)}$ to enter the radicand. When we use the square root or cube root option, the calculator considers the entire expression to be the radicand, so it is not necessary to close the parentheses. We do so, nevertheless, for completeness.

Higher order radical expressions can be entered using the $\sqrt[x]{}$ option from the MATH menu. When we use this option, we must enclose the radicand in parentheses if it contains more than one term. To enter $y_2 = \sqrt[4]{x - 1}$, position the cursor beside Y2 = on the equation-editor screen. Then press 4 to indicate that we are entering a fourth root. Next press $\boxed{\text{MATH}}$ 5 to select $\sqrt[x]{}$ from the MATH menu and finally press $\boxed{(}$ $\boxed{\text{X, T, }\Theta, n}$ $\boxed{-}$ $\boxed{1}$ $\boxed{)}$ to enter the radicand. Note that the grapher does not automatically supply a left parenthesis as it does when a square root or a cube root is selected.

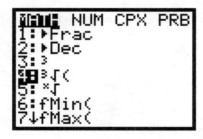   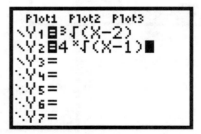

LINEAR REGRESSION

We can use the Linear Regression feature in the STAT CALC menu to fit a linear equation to a set of data.

Section 1.6, page 79 The following table lists data showing the price P of a one-day adult admission to Disney World for years since 1993.

Years, x (since 1993)	Price P of a One-Day Adult Admission to Disney World
0. 1993	$34.00
1. 1994	$36.00
2. 1995	$37.00
3. 1996	$40.81
4. 1997	$42.14
5. 1998	$44.52

(a) Fit a regression line to the data using the REGRESSION feature on a grapher.

(b) Graph the regression line with the scatterplot.

(c) Use the model to predict the price of a one-day adult admission in 2000 ($x = 7$).

(a) We will enter the data as ordered pairs on the STAT list editor screen. To clear any existing lists press $\boxed{\text{STAT}}$ 4 $\boxed{\text{2nd}}$ $\boxed{\text{L}_1}$ $\boxed{,}$ $\boxed{\text{2nd}}$ $\boxed{\text{L}_2}$ $\boxed{,}$ $\boxed{\text{2nd}}$ $\boxed{\text{L}_3}$ $\boxed{,}$ $\boxed{\text{2nd}}$ $\boxed{\text{L}_4}$ $\boxed{,}$ $\boxed{\text{2nd}}$ $\boxed{\text{L}_5}$ $\boxed{,}$ $\boxed{\text{2nd}}$ $\boxed{\text{L}_6}$ $\boxed{\text{ENTER}}$. (L_1 through L_6 are the second operations associated with the numeric keys 1 through 6.) The lists can also be cleared by first accessing the STAT list editor screen by pressing $\boxed{\text{STAT}}$ $\boxed{\text{ENTER}}$ or $\boxed{\text{STAT}}$ 1. These keystrokes display the STAT EDIT menu and then select the Edit option from that menu. Then, for each list that contains entries, use the arrow keys to move the cursor to highlight the name of the list at the top of the column and press $\boxed{\text{CLEAR}}$ $\boxed{\triangledown}$ or $\boxed{\text{CLEAR}}$ $\boxed{\text{ENTER}}$.

Once the lists are cleared, we can enter the data points. We will enter the number of years since 1993 in L_1 and the prices in L_2. Position the cursor at the top of column L_1, below the L_1 heading. To enter 0 press 0 $\boxed{\text{ENTER}}$. Continue typing the x-values 1 through 5, each followed by $\boxed{\text{ENTER}}$. The entries can be followed by $\boxed{\triangledown}$ rather than $\boxed{\text{ENTER}}$ if desired. Press $\boxed{\triangleright}$ to move to the top of column L_2. Type the prices 34.00, 36.00, and so on in succession, each followed by $\boxed{\text{ENTER}}$ or $\boxed{\triangledown}$. Note that the coordinates of each point must be in the same position in both lists.

```
L1        L2        L3        2
 0        34        ------
 1        36
 2        37
 3        40.81
 4        42.14
 5        44.52
------    ------
L2(7) =
```

The grapher's linear regression feature can be used to fit a linear equation to the data. Once the data has been entered in the lists, press $\boxed{\text{STAT}}$ $\boxed{\triangleright}$ 4 $\boxed{\text{ENTER}}$ to select LinReg($ax + b$) from the STAT CALC menu and to display the coefficients a and b of the regression equation $y = ax + b$.

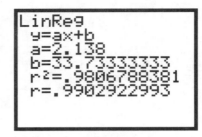

If the diagnostics have been turned on in your grapher, values for r^2 and r will also be displayed. These numbers indicate how well the regression line fits the data. For the remainder of this manual, regression will be done with the diagnostics turned off.

If you wish to select DiagnosticOn mode, press $\boxed{\text{2nd}}$ $\boxed{\text{CATALOG}}$ and use $\boxed{\triangledown}$ to position the triangular selection cursor beside DiagnosticOn. To alleviate the tedium of scrolling through many items to reach DiagnosticOn, press $\boxed{\text{D}}$ after pressing $\boxed{\text{2nd}}$ $\boxed{\text{CATALOG}}$ to move quickly to the first catalog item that begins with the letter D. (D is the ALPHA operation associated with the $\boxed{x^{-1}}$ key.) Then use $\boxed{\triangledown}$ to scroll to DiagnosticOn. Note that it is not necessary to press $\boxed{\text{ALPHA}}$ before $\boxed{\text{D}}$ when the catalog is displayed. Press $\boxed{\text{ENTER}}$ to paste this instruction to the home screen and then press $\boxed{\text{ENTER}}$ a second time to set the mode. To select DiagnosticOff mode, press $\boxed{\text{2nd}}$ $\boxed{\text{CATALOG}}$, position the selection cursor beside DiagnosticOff, press $\boxed{\text{ENTER}}$ to paste this instruction to the home screen, and then press $\boxed{\text{ENTER}}$ again to set this mode.

   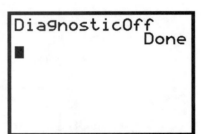

Immediately after the regression equation is found it can be copied to the equation-editor screen as Y_1. Note that any previous entry in Y_1 must have been cleared first. Press $\boxed{\text{Y} =}$ and position the cursor beside Y_1. Then press $\boxed{\text{VARS}}$ 5 $\boxed{\triangleright}$ $\boxed{\triangleright}$ 1. These keystrokes select Statistics from the VARS menu, then select the EQ (Equation) submenu, and finally select the RegEq (Regression Equation) from this submenu.

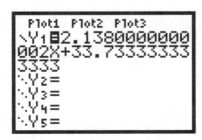

Before the regression equation is found, it is possible to select a y-variable to which it will be stored on the equation editor screen. After the data have been stored in the lists and the equation previously entered as Y_1 has been cleared, press STAT ▷ 4 VARS ▷ 1 1 ENTER . The coefficients of the regression equation will be displayed on the home screen, and the regression equation will also be stored as Y_1 on the equation-editor screen.

(b) To plot the data points, we turn on the STAT PLOT feature. To access the STAT PLOT screen, press 2nd STAT PLOT . (STAT PLOT is the second operation associated with the Y = key in the upper left-hand corner of the keypad.)

We will use Plot 1. Access it by highlighting 1 and pressing ENTER or simply by pressing 1. Now position the cursor over On and press ENTER to turn on Plot 1. The entries Type, Xlist, and Ylist should be as shown below. The last item, Mark, allows us to choose a box, a cross, or a dot for each point. Here we have selected a box. To select Type and Mark, position the cursor over the appropriate selection and press ENTER . Use the L_1 and L_2 keys (associated with the 1 and 2 numeric keys) to select Xlist and Ylist.

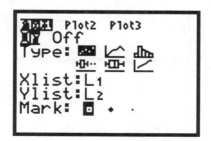

The plot can also be turned on from the "Y =" screen. Press Y = , the key at the top left-hand corner of the keypad, to go to this screen. Then, assuming that Plot 1 has not yet been turned on and that the desired settings are currently entered for Plot 1 on the STAT PLOT screen, position the cursor over Plot 1 and press ENTER . Plot 1 will now be highlighted.

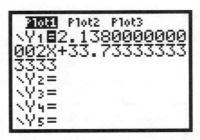

Now select a viewing window. The years range from 0 through 5 and the prices range from \$34.00 through \$44.52, so one good choice is $[-1, 6, 30, 50]$. To see the plotted points and the regression line, press GRAPH .

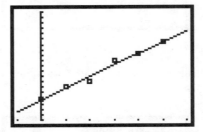

QUICK TIP: Instead of entering the window dimensions directly, we can press ZOOM 9 after entering the coordinates of the points in lists, turning on Plot 1, and selecting Type, Xlist, Ylist, and Mark. This activates the ZoomStat operation which automatically defines a viewing window that displays all of the points and also displays the graph.

(c) To predict the price of a one-day adult admission in 2000, evaluate the regression equation for $x = 7$. (2000 is 7 years after 1993.) Use any of the methods for evaluating a function presented earlier in this chapter. (See page 8.) We will use function notation on the home screen.

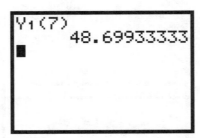

When $x = 7, y \approx 48.70$, so we predict that the price of a one-day adult admission to Disney World will be about \$48.70 in 2000.

POLYNOMIAL REGRESSION

The TI-83 and the TI-83 Plus have the capability to use regression to fit quadratic, cubic, and quartic functions to data.

Section 1.6, page 81 The following chart relates the number of live births to women of a particular age.

Age, x	Average Number of Live Births per 1000 Women
16	34
18.5	86.5
22	111.1
27	113.9
32	84.5
37	35.4
42	6.8

(a) Fit a quadratic function to the data using the REGRESSION feature on a grapher.

(b) Make a scatterplot of the data. Then graph the quadratic function with the scatterplot.

(c) Fit a cubic function to the data using the REGRESSION feature on a grapher.

(d) Make a scatterplot of the data. Then graph the cubic function with the scatterplot.

(e) Decide which function seems to fit the data better.

(f) Use the function from part (e) to estimate the average number of live births by women of ages 20 and 30.

(a) First enter the data with the ages in L$_1$ and the average number of live births per 1000 women in L$_2$. (See page 15 of this manual.) Then select quadratic regression, denoted QuadReg, from the STAT menu. Press $\boxed{\text{STAT}}$ $\boxed{\triangleright}$ 5 $\boxed{\text{ENTER}}$ to do this.

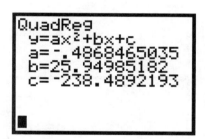

The grapher returns the coefficients of a quadratic function of the form $f(x) = ax^2 + bx + c$. Rounding the coefficients to two decimal places, we obtain the function $f(x) = -0.49x^2 + 25.95x - 238.49$.

(b) Graph the function along with the scatterplot as described on pages 17 and 18 of this manual.

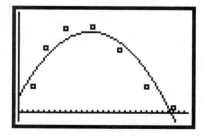

(c) Once the data are entered, fit a cubic function to it by pressing $\boxed{\text{STAT}}$ $\boxed{\triangleright}$ 6 $\boxed{\text{ENTER}}$. These keystrokes select the cubic regression feature, denoted CubicReg, from the STAT CALC menu and display the coefficients of a cubic function of the form $f(x) = ax^3 + bx^2 + cx + d$.

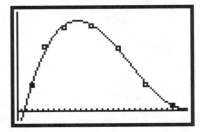

Rounding the coefficients to two decimal places, we have $f(x) = 0.03x^3 - 3.22x^2 + 101.18x - 886.93$.

(d) Graph the function along with the scatterplot as described on pages 17 and 18 of this manual.

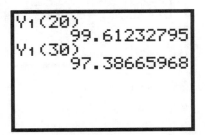

(e) The graph of the cubic function appears to fit the data better than the graph of the quadratic function.

(f) We can use any of the methods described earlier in this chapter to evaluate the function. We will use function notation on the home screen.

We estimate that there will be about 100 live births per 1000 20-year-old women and about 97 live births per 1000 30-year-old women.

A quartic function can be fit to data by selecting quartic regression, denoted QuartReg, from the STAT CALC menu. The grapher returns the coefficients of a function $f(x) = ax^4 + bx^3 + cx^2 + dx + e$.

Chapter 2
Differentiation

ZOOMING IN

The Zoom In operation from the ZOOM menu can be used to enlarge a portion of a graph.

Section 2.2, page 112 Verify graphically that $\lim_{x \to 0} \dfrac{\sqrt{x+1}-1}{x} = 0.5$.

First graph $y = \dfrac{\sqrt{x+1}-1}{x}$ in a window that shows the portion of the graph near $x = 0$. One good choice is $[-2, 5, -1, 2]$. Then press $\boxed{\text{TRACE}}$ and use the $\boxed{\triangleleft}$ and $\boxed{\triangleright}$ keys to move the trace cursor to a point on the curve near $x = 0$.

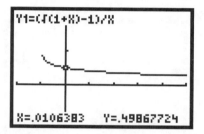

Now select the Zoom In operation from the ZOOM menu by pressing $\boxed{\text{ZOOM}}$ 2 $\boxed{\text{ENTER}}$. This enlarges the portion of the graph near $x = 0$. We can now press $\boxed{\text{TRACE}}$ again and use the $\boxed{\triangleleft}$ and $\boxed{\triangleright}$ keys to trace the curve near $x = 0$.

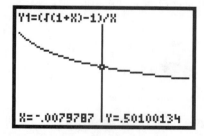 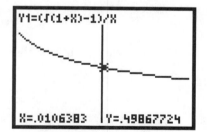

The Zoom In operation can be used as many times as desired in order to verify the result.

THE nDeriv OPERATION

The TI-83 and TI-83 Plus can be used to find the slope of a line tangent to a curve at a specific point. That is, they can find the derivative of a function $f(x)$ for a specific value of x.

Section 2.4, page 134 For the function $f(x) = x(100 - x)$, find $f'(70)$.

We will use the nDeriv (numerical derivative) operation. This is item 8 on the Math submenu of the MATH menu.

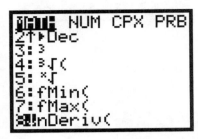

Select this operation by pressing $\boxed{\text{MATH}}$ 8 or by pressing $\boxed{\text{MATH}}$, using the $\boxed{\triangledown}$ key to highlight item 8, and then pressing $\boxed{\text{ENTER}}$. These keystrokes copy "nDeriv(" to the home screen. Now enter the function, the variable, and the value at which the derivative is to be evaluated, all separated by commas. Press $\boxed{\text{X,T,}\Theta,n}$ $\boxed{(}$ 1 0 0 $\boxed{-}$ $\boxed{\text{X,T,}\Theta,n}$ $\boxed{)}$ $\boxed{,}$ $\boxed{\text{X,T,}\Theta,n}$ $\boxed{,}$ 7 0 $\boxed{)}$ $\boxed{\text{ENTER}}$. Note that the grapher supplies a left parenthesis after "nDeriv" and we close the parentheses with a right parenthesis after entering 70. We see that $f'(70) = -40$.

THE TANGENT FEATURE

We can draw a line tangent to a curve at a given point using the Tangent feature from the DRAW submenu of the DRAW menu.

Section 2.4, page 134 Draw the line tangent to the graph of $f(x) = x(100 - x)$ at $x = 70$.

First graph $y_1 = x(100 - x)$ in a window that shows the portion of the curve near $x = 70$. One good choice is $[-10, 100, -10, 3000]$, Xscl = 10, Yscl = 1000. Be sure to clear any functions that were previously entered and turn off the plots.

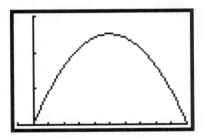

Now press $\boxed{\text{2nd}}$ $\boxed{\text{QUIT}}$ to go to the home screen. Here we will select the Tangent feature from the DRAW menu and instruct the grapher to draw the line tangent to the graph of y_1 at $x = 70$. To do this press $\boxed{\text{2nd}}$ $\boxed{\text{DRAW}}$ 5 $\boxed{\text{VARS}}$ $\boxed{\triangleright}$ 1 1 $\boxed{,}$ 7 0 $\boxed{)}$ $\boxed{\text{ENTER}}$. (DRAW is the second operation associated with the $\boxed{\text{PRGM}}$ key.) Note that the grapher supplies a left parenthesis after "Tangent" and we close the parentheses with a right parenthesis after entering 70.

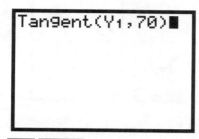
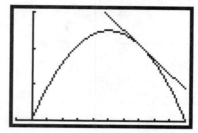

The keystrokes $\boxed{\text{2nd}}$ $\boxed{\text{DRAW}}$ 5 access the DRAW submenu of the DRAW menu and copy item 5, the Tangent feature, to the home screen. The same result can be achieved by pressing $\boxed{\text{2nd}}$ $\boxed{\text{DRAW}}$, using the $\boxed{\triangledown}$ key to highlight item 5, and then pressing $\boxed{\text{ENTER}}$. To clear the drawing from the GRAPH screen, use the ClrDraw (clear drawing) operation from the DRAW menu. Press $\boxed{\text{2nd}}$ $\boxed{\text{DRAW}}$ $\boxed{\text{ENTER}}$ or $\boxed{\text{2nd}}$ $\boxed{\text{DRAW}}$ 1.

See page 24 of this manual for a procedure that draws a tangent line directly from the GRAPH screen.

THE dy/dx OPERATION

We can find the derivative of a function at a specific point directly from the GRAPH screen.

Section 2.5, page 139 For the function $f(x) = x\sqrt{4 - x^2}$, find dy/dx at a specific point.

First graph $y = x\sqrt{4 - x^2}$. We will use the window $[-3, 3, -4, 4]$. Then select the dy/dx operation from the CALC menu by pressing $\boxed{\text{2nd}}$ $\boxed{\text{CALC}}$ 6 or by pressing $\boxed{\text{2nd}}$ $\boxed{\text{CALC}}$ to access the CALC menu, using the $\boxed{\triangledown}$ key to highlight item 6, dy/dx, and then pressing $\boxed{\text{ENTER}}$. We see the graph with a cursor positioned on it in the middle of the window which, in this case, is at $(0, 0)$.

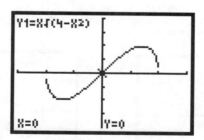

To find the value of dy/dx at a specific point either move the cursor to the desired point or key in its x-coordinate. For example, if we move the cursor to the point (1.212766, 1.9287169) and press $\boxed{\text{ENTER}}$ we find that $dy/dx = 0.66551339$ at this point.

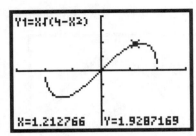   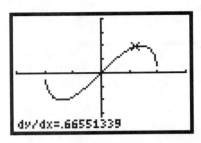

To find dy/dx for $x = 1$, select dy/dx from the CALC menu and then press 1 $\boxed{\text{ENTER}}$. We see that $dy/dx = 1.1547$ at this point.

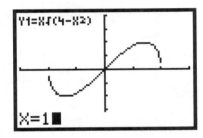   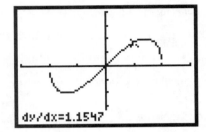

MORE ON TANGENT LINES

We can draw a line tangent to a curve at a specific point directly from the GRAPH screen.

Section 2.5, page 139 Draw the line tangent to the graph of $f(x) = x\sqrt{4 - x^2}$ at a specific point.

First graph $y = x\sqrt{4 - x^2}$ in an appropriate window such as $[-3, 3, -4, 4]$. Then press $\boxed{\text{TRACE}}$ and use the $\boxed{\triangleleft}$ and $\boxed{\triangleright}$ keys to move the cursor to the desired point. Now select the Tangent feature from the DRAW submenu of the DRAW menu and see the tangent line by pressing $\boxed{\text{2nd}}$ $\boxed{\text{DRAW}}$ 5 $\boxed{\text{ENTER}}$. The first coordinate of the point of tangency and the equation of the tangent line are also displayed.

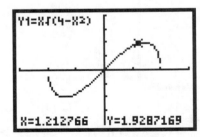   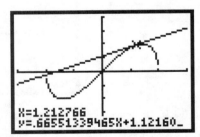

Use the ClrDraw (clear drawing) operation from the DRAW menu to clear the graph of the tangent line from the GRAPH screen. Press $\boxed{\text{2nd}}$ $\boxed{\text{DRAW}}$ $\boxed{\text{ENTER}}$ or $\boxed{\text{2nd}}$ $\boxed{\text{DRAW}}$ 1. To clear the graph of the tangent line when the home screen is displayed press $\boxed{\text{2nd}}$ $\boxed{\text{DRAW}}$ $\boxed{\text{ENTER}}$ $\boxed{\text{ENTER}}$ or $\boxed{\text{2nd}}$ $\boxed{\text{DRAW}}$ 1 $\boxed{\text{ENTER}}$.

Rather than using the TRACE feature to move the cursor to a point of tangency, we can also enter the x-coordinate of the point directly. For example, to graph the line tangent to the curve at $x = 1$, from the GRAPH screen press $\boxed{\text{2nd}}$ $\boxed{\text{DRAW}}$ 5 as before to select the Tangent feature. Then press 1 $\boxed{\text{ENTER}}$ to enter the x-coordinate, 1.

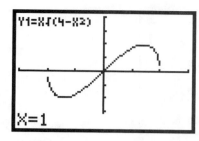 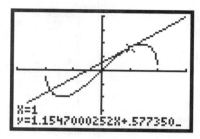

The graph of the tangent line can be cleared as described above.

If the arrow keys have been used to position the cursor at a point when using the dy/dx operation from the CALC menu, the tangent line at that point can be graphed from the GRAPH screen immediately after the value of dy/dx is displayed by pressing $\boxed{\text{2nd}}$ $\boxed{\text{DRAW}}$ 5 $\boxed{\text{ENTER}}$.

ENTERING THE SUM OF TWO FUNCTIONS

The Y-VARS submenu of the VARS menu can be used to enter a sum of functions.

Section 2.5, page 142 The Technology Connection on this page involves finding the derivative of a function $y_3 = y_1 + y_2$, where $y_1 = x(100 - x)$ and $y_2 = x\sqrt{100 - x^2}$. To enter y_3, first press $\boxed{\text{Y} =}$ to go to the equation-editor screen. Clear any entries present and turn off the plots. Then enter y_1 and y_2. To enter $y_3 = y_1 + y_2$, we select the functions y_1 and y_2 from the Y-VARS menu. Position the cursor beside "$Y_3 =$" and press $\boxed{\text{VARS}}$ $\boxed{\triangleright}$ 1 1 $\boxed{+}$ $\boxed{\text{VARS}}$ $\boxed{\triangleright}$ 1 2.

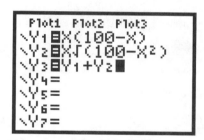

DESELECTING FUNCTIONS; GRAPH STYLES

Section 2.7, page 160 The Technology Connection on this page involves deselecting a function and also using different graph styles for two functions. First enter $y_1 = \dfrac{x^2 - 3x}{x - 1}$, $y_2 = \dfrac{x^2 - 2x + 3}{(x - 1)^2}$, and $y_3 = \text{nDeriv}(Y_1, x, x)$. Since we want to see only the graphs of y_2 and y_3, we will deselect y_1. To do this, position the cursor on the equals sign beside "Y_1" and press $\boxed{\text{ENTER}}$. Note that the equals sign is no longer highlighted. This indicates that y_1 has been deselected and, thus, its graph will not appear with the graphs of y_2 and y_3, the functions which continue to be selected.

To select y_1 again, position the cursor on the equals sign, beside "Y$_1$" and press ENTER. The equals sign will be highlighted now, indicating that the function is selected.

With y_1 deselected, we can graph y_2 and y_3 using different graph styles to determine whether the graphs coincide. When the grapher is in Connected mode, equations are graphed with a solid line.

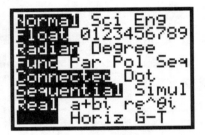

We will keep the solid line graph style for the graph of y_2 and select the path style for the graph of y_3. After the graph of y_2 is drawn, a circular cursor will trace the leading edge of the graph of y_3 and draw its path. (This assumes that the grapher is set in Sequential mode as shown above.)

To select the path style for y_3, position the cursor on the graph style icon to the left of "Y$_3$" on the "Y =" screen. Then press ENTER repeatedly until the path icon appears. The window below shows y_1 deselected, y_2 with the solid line graph style selected, and y_3 with the path style selected.

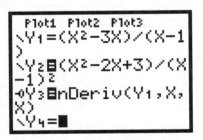

To see the graphs of y_2 and y_3 in the standard window, press ZOOM 6.

Chapter 3
Applications of Differentiation

THE MAXIMUM AND MINIMUM FEATURES

Section 3.1, page 195 Use the Maximum and Minimum features of a grapher to approximate the relative extrema of $f(x) = -0.4x^3 + 6.2x^2 - 11.3x - 54.8$.

First graph $y_1 = -0.4x^3 + 6.2x^2 - 11.3x - 54.8$ in a window that displays the relative extrema of the function. Trial and error reveals that one good choice is $[-10, 20, -100, 150]$, Xscl $= 5$, Yscl $= 50$. Observe that a relative maximum occurs near $x = 10$ and a relative minimum occurs near $x = 1$.

To find the relative maximum, first press $\boxed{\text{2nd}}$ $\boxed{\text{CALC}}$ 4 or $\boxed{\text{2nd}}$ $\boxed{\text{CALC}}$ $\boxed{\triangledown}$ $\boxed{\triangledown}$ $\boxed{\triangledown}$ $\boxed{\text{ENTER}}$ to select the Maximum feature from the CALC menu. We are prompted to select a left bound for the relative maximum. This means that we must choose an x-value that is to the left of the x-value of the point where the relative maximum occurs. This can be done by using the left- and right-arrow keys to move the cursor to a point to the left of the relative maximum or by keying in an appropriate value.

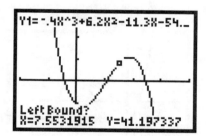

Once this is done, press $\boxed{\text{ENTER}}$. Now we are prompted to select a right bound. We move the cursor to a point to the right of the relative maximum or we key in an appropriate value.

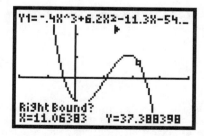

Press $\boxed{\text{ENTER}}$ again. Finally we are prompted to guess the x-value at which the relative maximum occurs. Move the cursor close to the relative maximum point or key in an x-value.

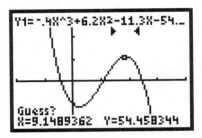

Press $\boxed{\text{ENTER}}$ a third time. We see that a relative maximum function value of approximately 54.61 occurs when $x \approx 9.32$.

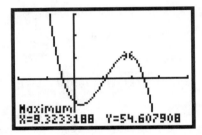

To find the relative minimum, select the Minimum feature from the CALC menu by pressing $\boxed{\text{2nd}}$ $\boxed{\text{CALC}}$ 3 or $\boxed{\text{2nd}}$ $\boxed{\text{CALC}}$ $\boxed{\triangledown}$ $\boxed{\triangledown}$ $\boxed{\text{ENTER}}$. Select left and right bounds for the relative minimum and guess the x-value at which it occurs as described above. We see that a relative minimum function value of approximately -60.30 occurs when $x \approx 1.01$.

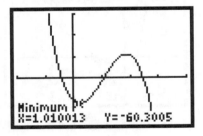

THE fMax AND fMin FEATURES

The fMax and fMin features can be used to calculate the x-values at which relative maximum and minimum values of a function occur over a specified closed interval.

Section 3.1, page 195 Use the fMax and fMin features of a grapher to approximate the relative extrema of $f(x) = -0.4x^3 + 6.2x^2 - 11.3x - 54.8$.

First enter the function as y_1 and graph it as described above. Observe that a relative maximum occurs in the interval $[5, 15]$. There are other intervals we could use. Keep in mind that the larger the interval, the longer it takes the grapher to return an x-value.

Now press $\boxed{\text{2nd}}$ $\boxed{\text{QUIT}}$ to go to the home screen and then press $\boxed{\text{MATH}}$ 7 to select the fMax feature from the MATH submenu of the MATH menu. Enter the name of the function, the variable, and the left and right endpoints of the interval on which the relative maximum occurs by pressing $\boxed{\text{VARS}}$ $\boxed{\triangleright}$ 1 1 $\boxed{,}$ $\boxed{\text{X,T,}\Theta,n}$ $\boxed{,}$ 5 $\boxed{,}$ 1 5 $\boxed{)}$. Press $\boxed{\text{ENTER}}$ to find that the relative maximum occurs when $x \approx 9.323320694$. To find the relative maximum value of the function, we evaluate the function for this value of x. Press $\boxed{\text{VARS}}$ $\boxed{\triangleright}$ 1 1 $\boxed{(}$ $\boxed{\text{2nd}}$ $\boxed{\text{ANS}}$ $\boxed{)}$ $\boxed{\text{ENTER}}$. (ANS is the second operation associated with the $\boxed{(-)}$ key.) The keystrokes $\boxed{\text{2nd}}$ $\boxed{\text{ANS}}$ cause the grapher to use the previous answer, 9.323320694, as the value for x in y_1.

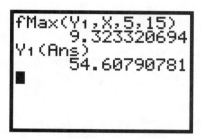

We also observe that a relative minimum occurs in the interval $[-5, 5]$. Again, there are other intervals we could choose. To find the relative minimum in this interval, first press $\boxed{\text{MATH}}$ 6 to select the fMin feature from the MATH submenu of the MATH menu. Then enter the name of the function, the variable, and the endpoints of the interval. Press $\boxed{\text{VARS}}$ $\boxed{\triangleright}$ 1 1 $\boxed{,}$ $\boxed{\text{X,T,}\Theta,n}$ $\boxed{,}$ $\boxed{(-)}$ 5 $\boxed{,}$ 5 $\boxed{)}$ $\boxed{\text{ENTER}}$. We see that a relative minimum function value occurs when $x \approx 1.010010343$. To find the relative minimum value of the function press $\boxed{\text{VARS}}$ $\boxed{\triangleright}$ 1 1 $\boxed{(}$ $\boxed{\text{2nd}}$ $\boxed{\text{ANS}}$ $\boxed{)}$ $\boxed{\text{ENTER}}$.

Chapter 4
Exponential and Logarithmic Functions

SOLVING EXPONENTIAL EQUATIONS

Section 4.2, page 307 Solve $e^t = 40$ using a grapher.

The Intersect feature is discussed on page 10 of this manual, and the Zero feature is discussed on page 11.

EXPONENTIAL REGRESSION

Section 4.3, page 323 To find an exponential equation that models a set of data, enter the data in lists as described on page 15 of this manual. Then select ExpReg (exponential regression) from the STAT CALC menu and find the regression equation by pressing $\boxed{\text{STAT}}$ $\boxed{\triangleright}$ 0 $\boxed{\text{ENTER}}$.

The grapher will return the values of a and b for an equation of the form $y = a \cdot b^x$. The function found can be evaluated for various values of x as described on page 8 of this manual.

Chapter 5
Integration

THE fnInt FEATURE

Definite integrals can be evaluated using the fnInt feature from the MATH submenu of the MATH menu.

Section 5.2, page 384 Evaluate $\int_{-1}^{2} (x^3 - 3x + 1)dx$ using the fnInt feature of a grapher.

First select the fnInt feature. This is item 9 on the MATH submenu of the MATH menu. Then enter the function, the variable, and the lower and upper limits of integration. Press $\boxed{\text{MATH}}$ $\boxed{9}$ $\boxed{\text{X,T,}\Theta,n}$ $\boxed{\wedge}$ $\boxed{3}$ $\boxed{-}$ $\boxed{3}$ $\boxed{\text{X,T,}\Theta,n}$ $\boxed{+}$ $\boxed{1}$ $\boxed{,}$ $\boxed{\text{X,T,}\Theta,n}$ $\boxed{,}$ $\boxed{(-)}$ $\boxed{1}$ $\boxed{,}$ $\boxed{2}$ $\boxed{)}$ $\boxed{\text{ENTER}}$. We find that $\int_{-1}^{2} (x^3 - 3x + 1)dx = 2.25$.

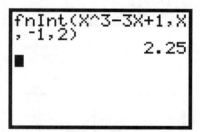

If the function has been entered on the Y = screen, as y_1 for instance, we can evaluate it by entering fnInt(Y$_1$, X, -1, 2) on the home screen. Press $\boxed{\text{MATH}}$ $\boxed{9}$ $\boxed{\text{VARS}}$ $\boxed{\triangleright}$ $\boxed{1}$ 1 $\boxed{,}$ $\boxed{\text{X,T,}\Theta,n}$ $\boxed{,}$ $\boxed{(-)}$ $\boxed{1}$ $\boxed{,}$ $\boxed{2}$ $\boxed{)}$ $\boxed{\text{ENTER}}$.

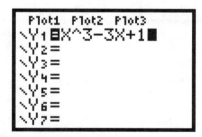
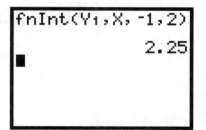

THE $\int f(x)dx$ FEATURE

Definite integrals can also be evaluated using the $\int f(x)dx$ feature from the CALC menu.

Section 5.2, page 384 Evaluate $\int_{-1}^{2} (x^3 - 3x + 1)dx$ using the $\int f(x)dx$ feature of a grapher.

First graph $y_1 = x^3 - 3x + 1$ in a window that contains the interval $[-1, 2]$. We will use $[-3, 3, -6, 6]$. Then select the $\int f(x)dx$ feature from the CALC menu by pressing $\boxed{\text{2nd}}$ $\boxed{\text{CALC}}$ 7. We are prompted to enter the lower limit of integration. Press $\boxed{(-)}$ 1.

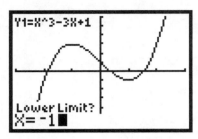

Then press ENTER . Now enter the upper limit of integration by pressing 2.

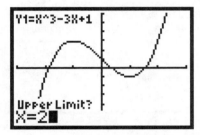

Press ENTER again. The grapher shades the area above and below the curve on $[-1, 2]$ and returns the value of the definite integral on this interval.

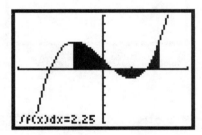

Chapter 6
Applications of Integration

STATISTICS

Section 6.8, page 475 The weights w of students in a calculus class are normally distributed with mean 150 lb and standard deviation 25 lb. Find the probability that a student's weight is from 160 lb to 180 lb.

It is not necessary to standardize the weights when the TI-83 or the TI-83 Plus is used. The grapher will graph the normal density function for the given mean and standard deviation, shade the area between 160 and 180, and express the probability as the area of the shaded region. A fair amount of trial and error might be required to find a suitable window. One good choice is $[0, 300, -0.002, 0.02]$, Xscl $= 50$, Yscl $= 0.01$. Enter these dimensions on the WINDOW screen. Clear any entries that are present on the equation-editor screen and be sure the plots are turned off. Then press 2nd DISTR ▷ 1 or 2nd DISTR ▷ ENTER to copy the ShadeNorm operation to the home screen. (DISTR is the second operation associated with the VARS key.) Then enter the left and right endpoints of the interval, the mean, and the standard deviation. To do this press 1 6 0 , 1 8 0 , 1 5 0 , 2 5) ENTER . The shaded area is 0.229509, so the probability is about 23%.

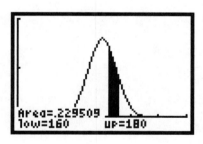

Chapter 7
Functions of Several Variables

PARTIAL DERIVATIVES

Section 7.2, page 505 Given the function $f(x, y) = 3x^3 y + 2xy$, use a grapher that finds derivatives of functions of one variable to find $f_x(-4, 1)$ and $f_y(2, 6)$.

To find $f_x(-4, 1)$, first find $f(x, 1)$:

$$f(x, y) = 3x^3 y + 2xy$$
$$f(x, 1) = 3x^3 \cdot 1 + 2x \cdot 1$$
$$= 3x^3 + 2x$$

Now we have a function of one variable, so we use the nDeriv operation or the dy/dx operation to find the value of the derivative of this function when $x = -4$. (See pages 21 and 23 of this manual for the procedures to follow.)

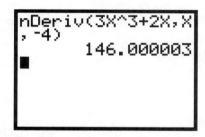 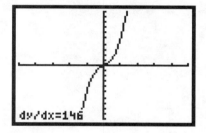

The procedures the grapher uses to calculate the derivative might not yield an exact answer. Note that the exact answer is 146, but the nDeriv operation yields 146.000003.

To find $f_y(2, 6)$, first find $f(2, y)$:

$$f(x, y) = 3x^3 y + 2xy$$
$$f(2, y) = 3 \cdot 2^3 y + 2 \cdot 2 \cdot y$$
$$= 24y + 4y = 28y$$

Now find the derivative of $f(y) = 28y$ when $y = 6$ using the nDeriv operation. Press $\boxed{\text{MATH}}$ 8 2 8 $\boxed{\text{ALPHA}}$ $\boxed{\text{Y}}$ $\boxed{,}$ $\boxed{\text{ALPHA}}$ $\boxed{\text{Y}}$ $\boxed{,}$ 6 $\boxed{)}$ $\boxed{\text{ENTER}}$. ($\boxed{\text{ALPHA}}$ is the green key in the left-hand column of the keypad and Y is the green ALPHA operation associated with the 1 numeric key.)

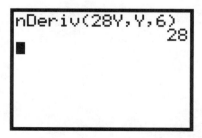

We can also replace y with x and find the derivative of $f(x) = 28x$ when $x = 6$ using the dy/dx operation.

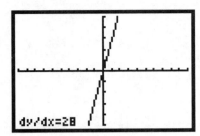

The TI-85
Graphics Calculator

Chapter 1
Functions, Graphs, and Models

GETTING STARTED

Press $\boxed{\text{ON}}$ to turn on the TI-85 graphing calculator. ($\boxed{\text{ON}}$ is the key at the bottom left-hand corner of the keypad.) You should see a blinking rectangle, or cursor, on the screen. If you do not see the cursor, try adjusting the display contrast. To do this, first press $\boxed{\text{2nd}}$. ($\boxed{\text{2nd}}$ is the yellow key in the left column of the keypad.) Then press and hold $\boxed{\triangle}$ to increase the contrast or $\boxed{\triangledown}$ to decrease the contrast. If the contrast needs to be adjusted further after the first adjustment, press $\boxed{\text{2nd}}$ again then then hold $\boxed{\triangle}$ or $\boxed{\triangledown}$ to increase or decrease the contrast, respectively.

To turn the grapher off, press $\boxed{\text{2nd}}$ $\boxed{\text{OFF}}$. (OFF is the second operation associated with the $\boxed{\text{ON}}$ key. In general, second operations are written in yellow above the keys on the keypad.) The grapher will turn itself off automatically after about five minutes without any activity.

Press $\boxed{\text{2nd}}$ $\boxed{\text{MODE}}$ to display the MODE settings. (MODE is the second operation associated with the $\boxed{\text{MORE}}$ key.) Initially you should select the settings on the left side of the display.

To change a setting on the Mode screen use $\boxed{\triangledown}$ or $\boxed{\triangle}$ to move the cursor to the line of that setting. Then use $\boxed{\triangleright}$ or $\boxed{\triangleleft}$ to move the blinking cursor to the desired setting and press $\boxed{\text{ENTER}}$. Press $\boxed{\text{CLEAR}}$ or $\boxed{\text{EXIT}}$ to leave the MODE screen. Pressing $\boxed{\text{CLEAR}}$ or $\boxed{\text{EXIT}}$ will take you to the home screen where computations are performed.

It will be helpful to read the Getting Started section of the Texas Instruments Guidebook that was packaged with your graphing calculator before proceeding.

USING A MENU

A menu is a list of options that appears when a key is pressed. Thus, multiple options, and sometimes multiple menus, may be accessed by pressing one key. For example, the following screen appears when $\boxed{\text{2nd}}$ $\boxed{\text{MATH}}$ is pressed. (MATH is the second operation associated with the $\boxed{\times}$ multiplication key.) We see several submenus at the bottom of the screen.

The $\boxed{\text{F1}}$ - $\boxed{\text{F5}}$ keys at the top of the keypad are used to select options from this menu. The arrow to the right of MISC indicates that there are more choices. They can be seen by pressing $\boxed{\text{MORE}}$.

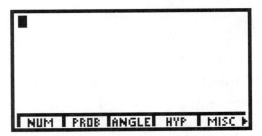

To choose the NUM submenu from the MATH menu press $\boxed{\text{F1}}$. (If you pressed $\boxed{\text{MORE}}$ to see the additional items on the MATH menu as described above, now press $\boxed{\text{MORE}}$ again to see the first five items on the menu. Then press $\boxed{\text{F1}}$ to choose NUM.) When NUM is chosen, the original submenus move up on the screen and the items on the NUM submenu appear at the bottom of the screen.

When two rows of options are displayed like this, the top row is accessed by pressing $\boxed{\text{2nd}}$ followed by one of the keys $\boxed{\text{F1}}$ - $\boxed{\text{F5}}$. These keystrokes access the second operations M1 - M5 associated with the $\boxed{\text{F1}}$ - $\boxed{\text{F5}}$ keys. The options on the bottom row are accessed by pressing one of the keys $\boxed{\text{F1}}$ - $\boxed{\text{F5}}$. Absolute value, denoted "abs," is selected from the NUM submenu and copied to the home screen, for instance, by pressing $\boxed{\text{F5}}$.

A menu can be removed from the screen by pressing $\boxed{\text{EXIT}}$. If both a menu and a submenu are displayed, press $\boxed{\text{EXIT}}$ once to remove the submenu and twice to remove both.

SETTING THE VIEWING RECTANGLE

Section 1.1, page 7 (Page numbers refer to pages in the textbook.) The viewing rectangle is the portion of the coordinate plane that appears on the grapher's screen. It is defined by the minimum and maximum values of x and y: xMin, xMax, yMin, and yMax. These are referred to as the RANGE variables. The notation [xMin, xMax, yMin, yMax] is used in the text to represent these settings or dimensions. For example, $[-12, 12, -8, 8]$ denotes a rectangle that displays the portion of the x-axis from -12 to 12 and the portion of the y-axis from -8 to 8. In addition, the distance between tick marks on the axes is defined by the settings xScl and yScl. In this manual xScl and yScl will be assumed to be 1 unless noted otherwise. The rectangle corresponding to the settings $[-20, 30, -12, 20]$, xScl = 5, yScl = 2, is shown below.

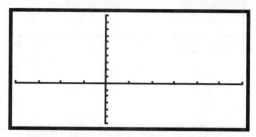

Press GRAPH F2 to display the RANGE screen and see the current settings on your grapher. The standard settings $[-10, 10, -10, 10]$, xScl = 1, yScl = 1, are shown below.

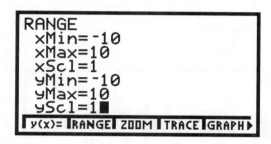

To change a setting, position the cursor beside the setting you wish to change and enter the new value. For example, to change from the standard settings to $[-20, \ 30, \ -12, \ 20]$, xScl = 5, yScl = 2, on the RANGE screen press (−) 2 0 ENTER 3 0 ENTER 5 ENTER (−) 1 2 ENTER 2 0 ENTER 2 ENTER. You must use the (−) key on the bottom of the keypad rather than the − key in the right-hand column to enter a negative number. (−) represents "the opposite of" or "the additive inverse of" whereas − is the key for the subtraction operation. The ▽ key may be used instead of ENTER after typing each setting. To see the viewing rectangle shown above, press F5 on the top row of the keypad.

QUICK TIP: To return quickly to the standard range settings $[-10, \ 10, \ -10, \ 10]$, xScl = 1, yScl = 1, press GRAPH F3 F4.

GRAPHS

After entering an equation and setting a viewing rectangle, you can view the graph of an equation.

Section 1.1, page 8 Graph $y = x^3 - 5x + 1$ using a graphing calculator.

Equations are entered on the y(x) =, or equation-editor, screen. Press GRAPH F1 to access this screen. If there is currently an expression displayed for y1, clear it by positioning the cursor beside "y1 =" and pressing CLEAR. Do the same for expressions that appear on all other lines by using ▽ to move to a line and then pressing CLEAR. Then use △ or ▽ to move the cursor beside "y1 =." Now press F1 ∧ 3 − 5 F1 + 1 to enter the right-hand side of the equation on the y(x) = screen. The keystroke F1 can be replaced with x-VAR. When the y(x) = screen is displayed, either keystroke will produce the variable x in an equation.

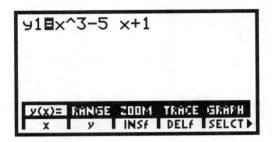

The standard $[-10, 10, -10, 10]$ viewing rectangle is a good choice for this graph. Either enter these dimensions on the RANGE screen and then press $\boxed{\text{F5}}$ to see the graph or, from the y(x) = screen, simply press $\boxed{\text{2nd}}$ $\boxed{\text{F3}}$ $\boxed{\text{F4}}$ or $\boxed{\text{GRAPH}}$ $\boxed{\text{F3}}$ $\boxed{\text{F4}}$ to select the standard rectangle and see the graph.

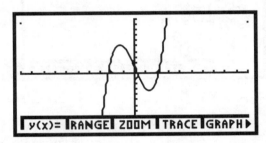

You can edit your entry if necessary. If, for instance, you pressed 6 instead of 5, use the $\boxed{\triangleleft}$ key to move the cursor to 6 and then press 5 to overwrite it. If you forgot to type the plus sign, move the cursor to 1, then press $\boxed{\text{2nd}}$ $\boxed{\text{INS}}$ $\boxed{+}$ to insert the plus sign before the 1. (INS is the second operation associated with the $\boxed{\text{DEL}}$ key.) You can continue to insert symbols immediately after the first insertion without pressing $\boxed{\text{2nd}}$ $\boxed{\text{INS}}$ again. If you typed 52 instead of 5, move the cursor to 2 and press $\boxed{\text{DEL}}$ to delete the 2.

An equation must be solved for y before it can be graphed on the TI-85.

Section 1.1, page 8 To graph $3x + 5y = 10$, first solve for y, obtaining $y = \dfrac{-3x + 10}{5}$. Then press $\boxed{\text{GRAPH}}$ $\boxed{\text{F1}}$ and clear any expressions that currently appear. Position the cursor beside y1 =. Now press $\boxed{(}$ $\boxed{(-)}$ 3 $\boxed{\text{F1}}$ $\boxed{+}$ 1 0 $\boxed{)}$ $\boxed{\div}$ 5 to enter the right-hand side of the equation. As before, $\boxed{\text{x-VAR}}$ can be used instead of $\boxed{\text{F1}}$. Note that without the parentheses the expression $-3x + \dfrac{10}{5}$ would have been entered.

Select a viewing rectangle and then, from the RANGE screen, press $\boxed{\text{F5}}$ to display the graph. You may change the viewing rectangle as desired to reveal more or less of the graph. The standard rectangle is shown here.

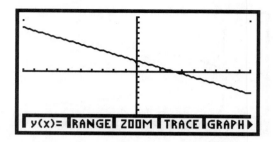

To graph $x = y^2$, first solve the equation for y : $y = \pm\sqrt{x}$. To obtain the entire graph of $x = y^2$, you must graph $y_1 = \sqrt{x}$ and $y_2 = -\sqrt{x}$ on the same screen. Press $\boxed{\text{GRAPH}}$ $\boxed{\text{F1}}$ and clear any expressions that currently appear. With the cursor beside y1 = press $\boxed{\text{2nd}}$ $\boxed{\sqrt{}}$ $\boxed{\text{F1}}$ or $\boxed{\text{2nd}}$ $\boxed{\sqrt{}}$ $\boxed{\text{x-VAR}}$. ($\boxed{\sqrt{}}$ is the second operation associated with the $\boxed{x^2}$ key.)

Now use $\boxed{\triangledown}$ to move the cursor beside y2 =. We will show two ways to enter $y_2 = -\sqrt{x}$. One is to enter the expression $-\sqrt{x}$ directly by pressing $\boxed{(-)}$ $\boxed{\text{2nd}}$ $\boxed{\sqrt{}}$ $\boxed{\text{F1}}$ or $\boxed{(-)}$ $\boxed{\text{2nd}}$ $\boxed{\sqrt{}}$ $\boxed{\text{x-VAR}}$.

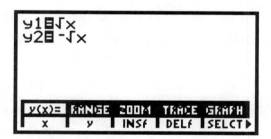

The other method of entering y_2 is based on the observation that $-\sqrt{x}$ is the opposite of the expression for y_1. That is, $y_2 = -y_1$. To enter this press $\boxed{(-)}$ $\boxed{\text{F2}}$ 1.

Select a viewing rectangle and press $\boxed{\text{F5}}$ to display the graph. The settings shown here are $[-2, 10, -5, 5]$.

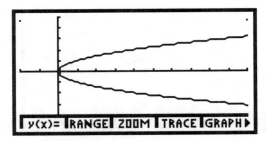

The top half is the graph of y_1, the bottom half is the graph of y_2, and together they yield the graph of $x = y^2$.

THE TABLE FEATURE

The TI-85 does not have a TABLE feature. However, there are several table programs for the TI-85 on the Texas Instruments web site, www.ti.com. These programs can be downloaded to your grapher.

GRAPHS AND FUNCTION VALUES

Section 1.2, page 21   There are several ways to evaluate a function using a grapher. Two of them are described here. Given the function $f(x) = 2x^2 + x$, we will find $f(-2)$. First press $\boxed{\text{GRAPH}}$ $\boxed{\text{F1}}$, clear all current entries, and enter the function as $y_1 = 2x^2 + x$.

We will find $f(-2)$ using the EVAL feature from the GRAPH menu. To do this, graph $y_1 = 2x^2 + x$ in a rectangle that includes the x-value -2. We will use the standard rectangle. Then, from the GRAPH screen, press $\boxed{\text{MORE}}$ $\boxed{\text{MORE}}$ $\boxed{\text{F1}}$ to select the EVAL feature. Now supply the desired x-value by pressing $\boxed{(-)}$ 2. Press $\boxed{\text{ENTER}}$ to see x = −2, y = 6 at the bottom of the screen. Thus $f(-2) = 6$.

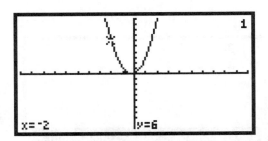

A function can also be evaluated on the home screen. With $y_1 = 2x^2 + x$ entered on the y(x) = screen, go to the home screen by pressing $\boxed{\text{2nd}}$ $\boxed{\text{QUIT}}$. (QUIT is the second operation associated with the $\boxed{\text{EXIT}}$ key.) Now enter $\boxed{(-)}$ 2 $\boxed{\text{STO}\triangleright}$ $\boxed{\text{x-VAR}}$ $\boxed{\text{ALPHA}}$ $\boxed{\text{2nd}}$ $\boxed{:}$ $\boxed{\text{2nd}}$ $\boxed{\text{alpha}}$ $\boxed{\text{Y}}$ 1 $\boxed{\text{ENTER}}$. After $\boxed{\text{STO}\triangleright}$ is pressed, the keyboard is set in ALPHA-lock. This means that the blue uppercase characters that appear on the six bottom rows of the keypad will appear when one of the corresponding keys is pressed. To cancel ALPHA-lock and and enter the remaining characters needed to evaluate this function, we press $\boxed{\text{ALPHA}}$. Then to enter the lowercase y we press $\boxed{\text{2nd}}$ $\boxed{\text{alpha}}$ $\boxed{\text{Y}}$. (: is the second operations associated with the $\boxed{.}$ key, alpha is the second operation associated with the blue $\boxed{\text{ALPHA}}$ key and $\boxed{\text{Y}}$ is the alphabetic operation associated with the 0 numeric key.)

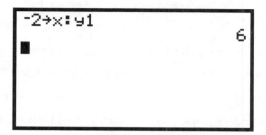

GRAPHING FUNCTIONS DEFINED PIECEWISE

Section 1.2, Example 9, page 22   Graph: $f(x) = \begin{cases} 4 \text{ for } x \le 0, \\ 4 - x^2 \text{ for } 0 < x \le 2, \\ 2x - 6 \text{ for } x > 2. \end{cases}$

We will enter the function using inequality symbols from the TEST menu. Press $\boxed{\text{GRAPH}}$ $\boxed{\text{F1}}$ to go to the equation-editor screen. Clear any entries that are present. Then position the cursor beside y1 = and press $\boxed{(}$ $\boxed{4}$ $\boxed{)}$ $\boxed{(}$ $\boxed{\text{x-VAR}}$ $\boxed{\text{2nd}}$ $\boxed{\text{TEST}}$ $\boxed{\text{F4}}$ $\boxed{0}$ $\boxed{)}$ $\boxed{+}$ $\boxed{(}$ $\boxed{4}$ $\boxed{-}$ $\boxed{\text{x-VAR}}$ $\boxed{x^2}$ $\boxed{)}$ $\boxed{(}$ $\boxed{0}$ $\boxed{\text{F2}}$ $\boxed{\text{x-VAR}}$ $\boxed{)}$ $\boxed{(}$ $\boxed{\text{x-VAR}}$ $\boxed{\text{F2}}$ $\boxed{2}$ $\boxed{)}$ $\boxed{+}$ $\boxed{(}$ $\boxed{2}$ $\boxed{\text{x-VAR}}$ $\boxed{-}$ $\boxed{6}$ $\boxed{)}$ $\boxed{(}$ $\boxed{\text{x-VAR}}$ $\boxed{\text{F3}}$ $\boxed{2}$ $\boxed{)}$. (TEST is the second operation associated with the 2 numeric key.) The keystrokes $\boxed{\text{2nd}}$ $\boxed{\text{TEST}}$ $\boxed{\text{F4}}$ open the TEST menu and select F4, the $\le$ symbol, from that menu. Similarly, with the TEST menu displayed, the keystrokes $\boxed{\text{F2}}$ and $\boxed{\text{F3}}$ select the symbols $<$ and $>$, respectively.

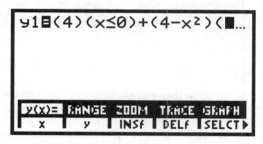

Now select DOT mode. Press $\boxed{\text{GRAPH}}$ $\boxed{\text{MORE}}$ $\boxed{\text{F3}}$ to display the FORMAT screen. Then use the arrow keys to highlight DrawDot and press $\boxed{\text{ENTER}}$.If this is not done, a vertical line that is not part of the graph will appear. Next choose and enter range dimensions and then press $\boxed{\text{F5}}$ to see the graph of the function. It is shown here in the rectangle $[-5, 5, -3, 6]$.

 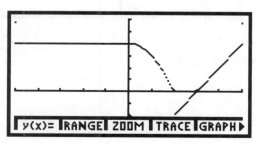

THE TRACE FEATURE

The TRACE feature can be used to display the coordinates of points on a graph.

Section 1.2, page 25 If you selected DrawDot mode for Example 9 above, return to DrawLine mode now. Press $\boxed{\text{GRAPH}}$ $\boxed{\text{MORE}}$ $\boxed{\text{F3}}$, highlight DrawLine, and the press $\boxed{\text{ENTER}}$. Now enter the function $f(x) = x^3 - 5x + 1$ (see page 43 of this manual) and graph it in the rectangle $[-5, 5, -10, 10]$. Then press $\boxed{\text{F4}}$ to select TRACE. A blinking cursor appears on the graph and the coordinates of the point at which it is positioned are displayed at the bottom of the screen. Use the $\boxed{\triangleleft}$ and $\boxed{\triangleright}$ keys to move the cursor along the graph to see the coordinates of other points.

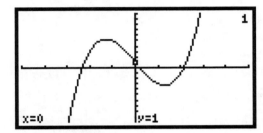 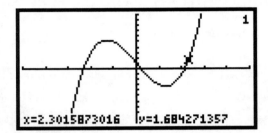

SQUARING THE VIEWING RECTANGLE

Section 1.4, page 41 In the standard rectangle, the distance between tick marks on the y-axis is about $3/5$ the distance between tick marks on the x-axis. It is often desirable to choose range dimensions for which these distances are the same, creating a "square" viewing rectangle. Any rectangle in which the ratio of the length of the y-axis to the length of the x-axis is $3/5$ will produce this effect. This can be accomplished by selecting dimensions for which yMax $-$ yMin $= \frac{3}{5}(\text{xMax} - \text{xMin})$.

The standard rectangle is shown on the left below and the square rectangle $[-10, 10, -6, 6]$ is shown on the right. Observe that the distance between tick marks appears to be the same on both axes when the square range variables are used.

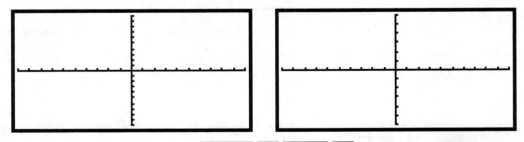

The rectangle can also be squared by pressing $\boxed{\text{GRAPH}}$ $\boxed{\text{F3}}$ $\boxed{\text{MORE}}$ $\boxed{\text{F2}}$ to select the ZSQR (Zoom Square) feature from the ZOOM menu. Starting with the standard rectangle and using the ZSQR feature, we get the dimensions and the rectangle shown below.

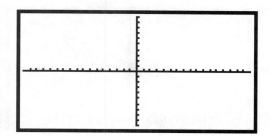

THE INTERSECT FEATURE

We can use the Intersect feature from the GRAPH MATH menu to solve equations.

Section 1.5, page 61 Solve the equation $x^3 = 3x + 1$ using the Intersect feature.

On the equation editor screen, clear any existing entries and then enter y1 $= x^3$ and y2 $= 3x + 1$. Now graph these equations in an appropriate window. One good choice is $[-3, 3, -10, 10]$. The solutions of the equation $x^3 = 3x + 1$ are the first coordinates of the points of intersection of these graphs. We will use the Intersect feature to find the leftmost point of intersection first. Press $\boxed{\text{MORE}}$ $\boxed{\text{F1}}$ $\boxed{\text{MORE}}$ $\boxed{\text{F5}}$ to select ISECT (Intersect) from the GRAPH MATH menu. A cursor appears on the graph of y1. Use the left and/or right arrow keys to move the cursor near the point you wish to find. Then press $\boxed{\text{ENTER}}$ to indicate that this is the first curve involved in the intersection. The cursor will move to the graph of y2. Press $\boxed{\text{ENTER}}$ to indicate that this is the second curve. We identify the curves for the grapher since we could have as many as ten graphs on the screen at once. After we identify the second curve, the coordinates of the point of intersection appear at the bottom of the screen.

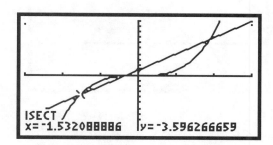

We see that, at the leftmost point of intersection, $x \approx -1.53$, so one solution of the equation is approximately -1.53. Repeat this process two times to find the coordinates of the other two points of intersection. As we see in the windows on the next page, the other two solutions of the equation are approximately -0.35 and 1.88.

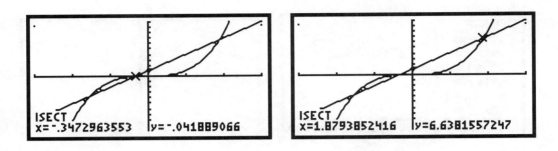

THE ROOT FEATURE

When an equation is expressed in the form $f(x) = 0$, it can be solved using the ROOT feature from the GRAPH MATH menu.

Section 1.5, page 61 Solve the equation $x^3 = 3x + 1$ using the ROOT feature.

First subtract $3x$ and 1 on both sides of the equation to obtain an equivalent equation with 0 on one side. We have $x^3 - 3x - 1 = 0$. The solutions of the equation $x^3 = 3x + 1$ are the values of x for which the function $f(x) = x^3 - 3x - 1$ is equal to 0. We can use the ROOT feature to find these values, or zeros.

On the equation-editor screen, clear any existing entries and then enter $y_1 = x^3 - 3x - 1$. Now graph the function in a viewing rectangle that shows the x-intercepts clearly. One good choice is $[-3, 3, -5, 8]$. We see that the function has three zeros. They appear to be about -1.5, -0.5, and 2.

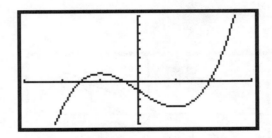

We will find the zero near -1.5 first. Note that in the window below we have cleared the GRAPH menu by pressing CLEAR . Press GRAPH to display the menu again. Then press MORE F1 F3 to select the ROOT feature from the GRAPH MATH menu. Use the left and/or right arrow key to move the cursor near the leftmost zero and press ENTER .

We see that -1.53 is a zero of the function. The procedure the ROOT operation uses to find the zeros might not give the value of y exactly. Note here, for example, that the y-value shown is 2E$-$13, or 2×10^{-13}, or 0.0000000000002 wish is very close to zero.

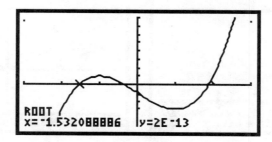

Select ROOT from the GRAPH MATH menu a second time to find the zero near -0.5 and a third time to find the zero near 2. We see that the other two zeros are approximately -0.35 and 1.88.

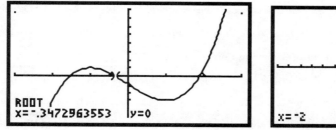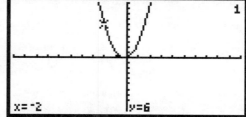

ABSOLUTE-VALUE FUNCTIONS

We can use the absolute-value option to perform computations involving absolute value and to graph absolute-value functions.

Section 1.5, page 65   Graph $f(x) = |x|$.

The absolute-value option can be accessed from either the MATH NUM (Number) menu or from the catalog. Before either option is chosen, first press $\boxed{\text{GRAPH}}$ $\boxed{\text{F1}}$ to go to the equation-editor screen and then clear any existing entries. Now position the cursor beside y1 = and enter $|x|$ as abs x.

To do this using the MATH NUM menu, first press $\boxed{\text{2nd}}$ $\boxed{\text{MATH}}$ $\boxed{\text{F1}}$ $\boxed{\text{F5}}$ to copy "abs" to the equation-editor screen. (MATH is the second operation associated with the $\boxed{\times}$ multiplication key.) Then press $\boxed{\text{x-VAR}}$. To select "abs" from the catalog, press $\boxed{\text{2nd}}$ $\boxed{\text{CATALOG}}$ $\boxed{\text{F1}}$. (CATALOG is the second operation associated with the $\boxed{\text{CUSTOM}}$ key.) If the cursor is positioned beside "abs," press $\boxed{\text{ENTER}}$ to copy "abs" to the equation-editor screen. If not, press $\boxed{\text{A}}$ to move the cursor beside the first catalog entry that begins with A. This is "abs," so we now press $\boxed{\text{ENTER}}$ to copy "abs" to the equation-editor screen. ("A" is the alpha operation associated with the $\boxed{\text{LOG}}$ key.) Then press $\boxed{\text{x-VAR}}$. In either case, choose an appropriate viewing rectangle and graph the function. To use the standard viewing rectangle, press $\boxed{\text{GRAPH}}$ $\boxed{\text{F3}}$ $\boxed{\text{F4}}$.

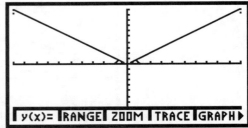

GRAPHING RADICAL FUNCTIONS

There are various way to enter radical expressions on the TI-85.

Section 1.5, page 67 We discussed entering an expression containing a square root on page 45 of this manual. If the radicand has more than one term, the entire radicand must be enclosed in parentheses. To enter $y_1 = \sqrt{x+2}$, for example, position the cursor beside y1 = on the equation editor screen and press $\boxed{\text{2nd}}$ $\boxed{\sqrt{}}$ $\boxed{(}$ $\boxed{\text{x-VAR}}$ $\boxed{+}$ $\boxed{2}$ $\boxed{)}$.

Higher order radical expressions can be entered using the $\sqrt[x]{}$ option from the MATH MISC menu. We must enclose the radicand in parentheses if it contains more than one term. To enter $y_2 = \sqrt[3]{x-2}$, position the cursor beside y2 = on the equation-editor screen. Then press 3 to indicate that we are entering a cube root. Next press $\boxed{\text{2nd}}$ $\boxed{\text{MATH}}$ $\boxed{\text{F5}}$ $\boxed{\text{MORE}}$ $\boxed{\text{F4}}$ to select $\sqrt[x]{}$. Finally press $\boxed{(}$ $\boxed{\text{x-VAR}}$ $\boxed{-}$ $\boxed{2}$ $\boxed{)}$ to enter the radicand.

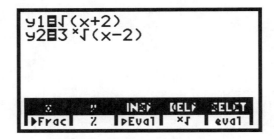

LINEAR REGRESSION

We can use the Linear Regression feature in the STAT CALC menu to fit a linear equation to a set of data.

Section 1.6, page 79 The following table lists data showing the price P of a one-day adult admission to Disney World for years since 1993.

Years, x (since 1993)	Price P of a One-Day Adult Admission to Disney World
0. 1993	$34.00
1. 1994	$36.00
2. 1995	$37.00
3. 1996	$40.81
4. 1997	$42.14
5. 1998	$44.52

(a) Fit a regression line to the data using the REGRESSION feature on a grapher.

(b) Graph the regression line with the scatterplot.

(c) Use the model to predict the price of a one-day adult admission in 2000 ($x = 7$).

(a) We will enter the data as ordered pairs on the STAT list editor screen. To clear any existing lists press STAT F2 ENTER ENTER F5.

Once the lists are cleared, we can enter the data points. We will enter the number of years since 1993 in xStat and the prices in yStat. Position the cursor to the right of x1 =. To enter the data point (0, 34.00) press 0 ENTER 3 4 ENTER. The entries can be followed by ▽ rather than ENTER if desired. Now enter the data point (1, 36.00) by pressing 1 ENTER 3 6 ENTER. Continue in this manner until all the data points are entered.

The grapher's linear regression feature can be used to fit a linear equation to the data. Once the data have been entered in the lists, press 2nd F1 to display the CALC menu. Then press ENTER ENTER F2 to select LINR (linear regression) from the STAT CALC menu. The coefficients a and b of the regression equation $y = a + bx$ are displayed along with corr, the coefficient of correlation, and the number n of data points used.

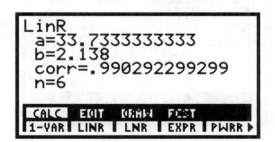

Immediately after the regression equation is found it can be copied to the equation-editor screen as y1. Note that any previous entry in y1 must have been cleared first. Press GRAPH F1 and position the cursor beside y1 =. Then press 2nd VARS MORE MORE F3. (VARS is the second operation associated with the 3 numeric key.) These keystrokes select Statistics from the VARS menu. Use ▽ to move the selection cursor beside RegEq and then press ENTER to enter this equation as y1.

(b) To plot the data points, we turn on the STAT PLOT feature. First select a viewing rectangle. The years range from 0 through 5 and the prices range from \$34.00 through \$44.52, so one good choice is $[-1, 6, 30, 50]$. Then select SCAT (scatterplot) from the STAT DRAW menu by pressing $\boxed{\text{STAT}}$ $\boxed{\text{F3}}$ $\boxed{\text{F2}}$.

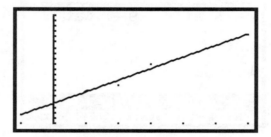

(c) To predict the price of a one-day adult admission in 2000, evaluate the regression equation for $x = 7$. (2000 is 7 years after 1993.) We can use the FCST (Forecast) operation fro the STAT menu to do this. Press $\boxed{\text{2nd}}$ $\boxed{\text{QUIT}}$ to go to the home screen. Press $\boxed{\text{STAT}}$ $\boxed{\text{F4}}$ to display the forecasting screen. Then press 7 $\boxed{\text{ENTER}}$ $\boxed{\text{F5}}$ (SOLVE).

When $x = 7, y \approx 48.70$, so we predict that the price of a one-day adult admission to Disney World will be about \$48.70 in 2000.

We could also use one of the methods for evaluating a function presented earlier in this chapter to make this prediction. (See page 46 of this manual.)

POLYNOMIAL REGRESSION

The TI-85 has the capability to use regression to fit quadratic, cubic, and quartic functions to data.

Section 1.6, page 81 The following chart relates the number of live births to women of a particular age.

Age, x	Average Number of Live Births per 1000 Women
16	34
18.5	86.5
22	111.1
27	113.9
32	84.5
37	35.4
42	6.8

(a) Fit a quadratic function to the data using the REGRESSION feature on a grapher.

(b) Make a scatterplot of the data. Then graph the quadratic function with the scatterplot.

(c) Fit a cubic function to the data using the REGRESSION feature on a grapher.

(d) Make a scatterplot of the data. Then graph the cubic function with the scatterplot.

(e) Decide which function seems to fit the data better.

(f) Use the function from part (e) to estimate the average number of live births by women of ages 20 and 30.

(a) First clear the data lists and enter the new data with the ages in xStat and the average number of live births per 1000 women in yStat. (See page 53 of this manual.) Then select quadratic regression, denoted P2Reg, from the STAT CALC menu. Press $\boxed{\text{2nd}}$ $\boxed{\text{F1}}$ $\boxed{\text{ENTER}}$ $\boxed{\text{ENTER}}$ $\boxed{\text{MORE}}$ $\boxed{\text{F1}}$ to do this.

The grapher returns a list PRegC = $\{a \ b \ c\}$ of the coefficients of a quadratic function of the form $f(x) = ax^2 + bx + c$. Rounding the coefficients to two decimal places, we obtain the function $f(x) = -0.49x^2 + 25.95x - 238.49$.

(b) Graph the function along with the scatterplot as described on page 54 of this manual. The viewing rectangle shown here is $[10, 50, 0, 125]$, xScl $= 10$, yScl $= 25$.

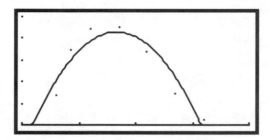

(c) Once the data are entered, fit a cubic function to them from the EDIT screen by pressing ⌐2nd⌐ ⌐F1⌐ ⌐ENTER⌐ ⌐ENTER⌐ ⌐MORE⌐ ⌐F2⌐. From the home screen press ⌐STAT⌐ ⌐F1⌐ ⌐ENTER⌐ ⌐ENTER⌐ ⌐MORE⌐ ⌐F2⌐. These keystrokes select the cubic regression feature, denoted P3Reg, from the STAT CALC menu and display a list PRegC $=\{a \quad b \quad c \quad d\}$ of the coefficients of a cubic function of the form $f(x) = ax^3 + bx^2 + cx + d$.

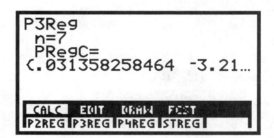

Rounding the coefficients to two decimal places, we have $f(x) = 0.03x^3 - 3.22x^2 + 101.18x - 886.93$.

(d) Graph the function along with the scatterplot as described on page 54 of this manual.

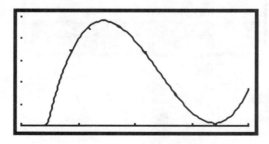

(e) The graph of the cubic function appears to fit the data better than the graph of the quadratic function.

(f) We can use any of the methods described earlier in this chapter to evaluate the function. We will use Forecast feature from the STAT menu. (See page 54.)

 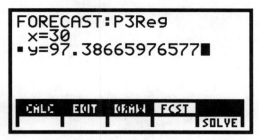

We estimate that there will be about 100 live births per 1000 20-year-old women and about 97 live births per 1000 30-year-old women.

A quartic function can be fit to data by selecting quartic regression, denoted P4Reg, from the STAT CALC menu. The grapher returns a list PRegC = $\{a \quad b \quad c \quad d \quad e\}$ of the coefficients of a function $f(x) = ax^4 + bx^3 + cx^2 + dx + e$.

Chapter 2
Differentiation

ZOOMING IN

The ZIN (Zoom In) operation from the GRAPH ZOOM menu can be used to enlarge a portion of a graph.

Section 2.2, page 112 Verify graphically that $\lim\limits_{x \to 0} \dfrac{\sqrt{x+1}-1}{x} = 0.5$.

First graph $y = \dfrac{\sqrt{x+1}-1}{x}$ in a rectangle that shows the portion of the graph near $x = 0$. One good choice is $[-2, 5, -1, 2]$. Note that the radicand, $x + 1$, must be enclosed in parentheses. That is, we enter the equation as $y = (\sqrt{(x+1)} - 1)/x$. Without parentheses the grapher would read the expression as $y = \dfrac{\sqrt{x}+1-1}{x}$. Then press $\boxed{\text{F3}}$ $\boxed{\text{F2}}$ to select ZIN and use the $\boxed{\triangleleft}$ and $\boxed{\triangleright}$ keys to move the cursor to a point on the curve near $x = 0$.

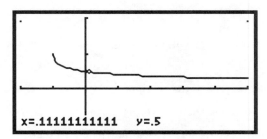

Now press $\boxed{\text{ENTER}}$. This enlarges the portion of the graph near $x = 0$. We can now press $\boxed{\text{GRAPH}}$ $\boxed{\text{F4}}$ and use the $\boxed{\triangleleft}$ and $\boxed{\triangleright}$ keys to trace the curve near $x = 0$.

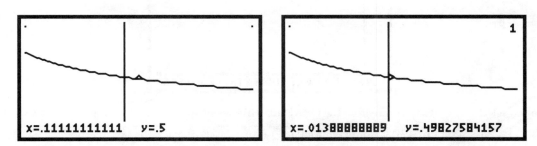

The ZIN operation can be used as many times as desired in order to verify the result.

THE nDer OPERATION

The TI-85 can be used to find the slope of a line tangent to a curve at a specific point. That is, it can find the derivative of a function $f(x)$ for a specific value of x.

Section 2.4, page 134 For the function $f(x) = x(100 - x)$, find $f'(70)$.

We will use the nDer (numerical derivative) operation. Select this operation by pressing $\boxed{\text{2nd}}$ $\boxed{\text{CALC}}$ $\boxed{\text{F2}}$. (CALC is the second operation associated with the $\boxed{\div}$ key.) These keystrokes copy "nDer(" to the home screen. Now enter the function, the variable, and the value at which the derivative is to be evaluated, all separated by commas. Press $\boxed{\text{x-VAR}}$ $\boxed{(}$ $\boxed{1}$ $\boxed{0}$ $\boxed{0}$ $\boxed{-}$ $\boxed{\text{x-VAR}}$ $\boxed{)}$ $\boxed{,}$ $\boxed{\text{x-VAR}}$ $\boxed{,}$ $\boxed{7}$ $\boxed{0}$ $\boxed{)}$ $\boxed{\text{ENTER}}$. Note that the grapher supplies a left parenthesis after "nDer" and we close the parentheses with a right parenthesis after entering 70. We see that $f'(70) = -40$.

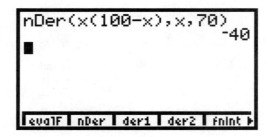

THE TANGENT FEATURE

We can draw a line tangent to a curve at a given point using the Tangent feature from the DRAW submenu of the GRAPH menu.

Section 2.4, page 134 Draw the line tangent to the graph of $f(x) = x(100 - x)$ at $x = 70$.

First graph $y_1 = x(100 - x)$ in a rectangle that shows the portion of the curve near $x = 70$. One good choice is $[-10, 100, -10, 3000]$, xScl = 10, yScl = 1000. Be sure to clear any functions that were previously entered.

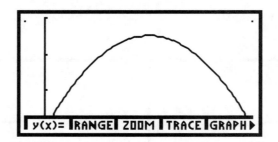

Here we will select the Tangent feature from the GRAPH DRAW menu and instruct the grapher to draw the line tangent to the graph of y_1 at $x = 70$. To do this press $\boxed{\text{MORE}}$ $\boxed{\text{F2}}$ $\boxed{\text{MORE}}$ $\boxed{\text{MORE}}$ $\boxed{\text{F1}}$ $\boxed{\text{2nd}}$ $\boxed{\text{alpha}}$ $\boxed{\text{Y}}$ $\boxed{1}$ $\boxed{,}$ $\boxed{7}$ $\boxed{0}$ $\boxed{)}$ $\boxed{\text{ENTER}}$. Note that the grapher supplies a left parenthesis after "TanLn" and we close the parentheses with a right parenthesis after entering 70.

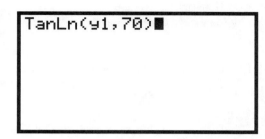 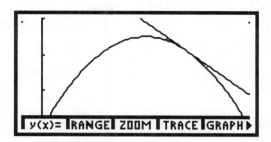

To clear the graph of the tangent line from the screen use the CLDRAW (clear drawing) operation from the GRAPH DRAW menu. Press $\boxed{\text{MORE}}$ $\boxed{\text{F2}}$ $\boxed{\text{MORE}}$ $\boxed{\text{F5}}$.

See page 62 of this manual for a procedure that draws a tangent line directly from the GRAPH screen.

THE dy/dx OPERATION

We can find the derivative of a function at a specific point directly from the GRAPH screen.

Section 2.5, page 139 For the function $f(x) = x\sqrt{4 - x^2}$, find dy/dx at a specific point.

First graph $y = x\sqrt{(4 - x^2)}$. We will use the rectangle $[-3, 3, -4, 4]$. Then select the dy/dx operation from the GRAPH MATH menu by pressing $\boxed{\text{MORE}}$ $\boxed{\text{F1}}$ $\boxed{\text{F4}}$. We see the graph with a cursor positioned on it in the middle of the viewing rectangle which, in this case, is at $(0, 0)$.

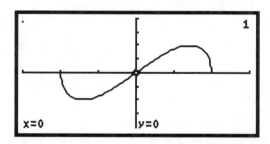

To find the value of dy/dx at a specific point move the cursor to the desired point. For example, if we move the cursor to the point $(1.2380952381, 1.9446847395)$ and press $\boxed{\text{ENTER}}$ we find that $dy/dx = 0.5947897471$ at this point.

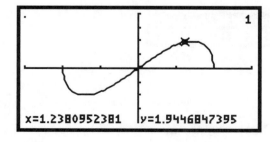 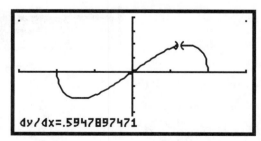

MORE ON TANGENT LINES

We can draw a line tangent to a curve at a specific point directly from the GRAPH screen.

Section 2.5, page 139 Draw the line tangent to the graph of $f(x) = x\sqrt{4 - x^2}$ at a specific point.

First graph $y = x\sqrt{(4 - x^2)}$ in an appropriate viewing rectangle such as $[-3, 3, -4, 4]$. Select the TANLN feature from the MATH submenu of the GRAPH menu and see the tangent line by first pressing $\boxed{\text{MORE}}$ $\boxed{\text{F1}}$ $\boxed{\text{MORE}}$ $\boxed{\text{MORE}}$ $\boxed{\text{F3}}$. Then move the cursor to the desired point and press $\boxed{\text{ENTER}}$. The value of the derivative at the point of tangency is displayed.

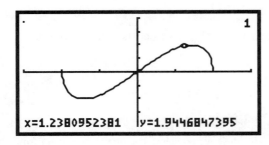 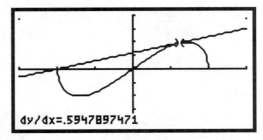

Use the CLDRAW (clear drawing) operation from the GRAPH DRAW menu to clear the graph of the tangent line from the GRAPH screen. Press $\boxed{\text{GRAPH}}$ $\boxed{\text{MORE}}$ $\boxed{\text{F2}}$ $\boxed{\text{MORE}}$ $\boxed{\text{F5}}$.

ENTERING THE SUM OF TWO FUNCTIONS

Section 2.5, page 142 The Technology Connection on this page involves finding the derivative of a function $y_3 = y_1 + y_2$, where $y_1 = x(100 - x)$ and $y_2 = x\sqrt{100 - x^2}$. To enter y_3, first press $\boxed{\text{GRAPH}}$ $\boxed{\text{F1}}$ to go to the equation-editor screen. Clear any entries present. Then enter y_1 and y_2. To enter $y_3 = y_1 + y_2$, position the cursor beside y3 = and press $\boxed{\text{F2}}$ 1 $\boxed{+}$ $\boxed{\text{F2}}$ 2 or $\boxed{\text{2nd}}$ $\boxed{\text{alpha}}$ $\boxed{\text{Y}}$ 1 $\boxed{+}$ $\boxed{\text{2nd}}$ $\boxed{\text{alpha}}$ $\boxed{\text{Y}}$ 2.

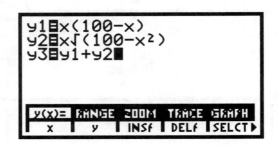

DESELECTING FUNCTIONS

Section 2.7, page 160 The Technology Connection on this page involves deselecting a function. First enter $y_1 = \dfrac{x^2 - 3x}{x - 1}$, $y_2 = \dfrac{x^2 - 2x + 3}{(x - 1)^2}$, and $y_3 = \text{nDer}(y1, x, x)$. Since we want to see only the graphs of y_2 and y_3, we will deselect y_1. To do this, position the cursor anywhere in the function y1 and press $\boxed{\text{F5}}$ (SELCT). Note that the equals sign is no longer highlighted. This indicates that y_1 has been deselected and, thus, its graph will not appear with the graphs of y_2 and y_3, the functions which continue to be selected.

To select y_1 again, position the cursor anywhere in the function y1 and press F5 . The equals sign will be highlighted now, indicating that the function is selected.

Chapter 3
Applications of Differentiation

THE FMAX AND FMIN FEATURES

Section 3.1, page 195 Use the FMAX and FMIN features of a grapher to approximate the relative extrema of $f(x) = -0.4x^3 + 6.2x^2 - 11.3x - 54.8$.

First graph $y_1 = -0.4x^3 + 6.2x^2 - 11.3x - 54.8$ in a viewing rectangle that displays the relative extrema of the function. Trial and error reveals that one good choice is $[-10, 20, -100, 150]$, xScl $= 5$, yScl $= 50$. Observe that a relative maximum occurs near $x = 10$ and a relative minimum occurs near $x = 1$.

To find the relative maximum, first press $\boxed{\text{MORE}}$ $\boxed{\text{F1}}$ $\boxed{\text{MORE}}$ $\boxed{\text{F2}}$ to select the FMAX feature from the GRAPH MATH menu. Move the cursor near the relative maximum point and press $\boxed{\text{ENTER}}$. We see that a relative maximum function value of approximately 54.61 occurs when $x \approx 9.32$.

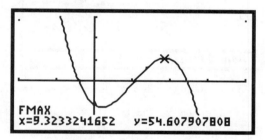

To find the relative minimum, select the FMIN feature from the GRAPH MATH menu by pressing $\boxed{\text{GRAPH}}$ $\boxed{\text{MORE}}$ $\boxed{\text{F1}}$ $\boxed{\text{MORE}}$ $\boxed{\text{F1}}$. Move the cursor near the relative minimum and press $\boxed{\text{ENTER}}$. We see that a relative minimum function value of approximately -60.30 occurs when $x \approx 1.01$.

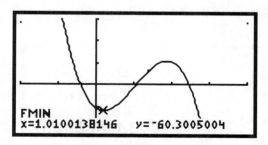

THE fMax AND fMin FEATURES

The fMax and fMin features can be used to calculate the x-values at which relative maximum and minimum values of a function occur over a specified closed interval.

Section 3.1, page 195　Use the fMax and fMin features of a grapher to approximate the relative extrema of $f(x) = -0.4x^3 + 6.2x^2 - 11.3x - 54.8$.

First enter the function as y_1 and graph it as described above. Observe that a relative maximum occurs in the interval $[5, 15]$. There are other intervals we could use. Keep in mind that the larger the interval, the longer it takes the grapher to return an x-value.

Now press $\boxed{\text{2nd}}$ $\boxed{\text{QUIT}}$ to go to the home screen and then press $\boxed{\text{2nd}}$ $\boxed{\text{CALC}}$ $\boxed{\text{MORE}}$ $\boxed{\text{F2}}$ to select the fMax feature from the CALC menu. Enter the name of the function, the variable, and the left and right endpoints of the interval on which the relative maximum occurs by pressing $\boxed{\text{2nd}}$ $\boxed{\text{alpha}}$ $\boxed{\text{Y}}$ $\boxed{1}$ $\boxed{,}$ $\boxed{\text{x-VAR}}$ $\boxed{,}$ $\boxed{5}$ $\boxed{,}$ $\boxed{1}$ $\boxed{5}$ $\boxed{)}$. Press $\boxed{\text{ENTER}}$ to find that the relative maximum occurs when $x \approx 9.32332069352$. To find the relative maximum value of the function, we evaluate the function for this value of x. Press $\boxed{\text{STO▷}}$ $\boxed{\text{x-VAR}}$ $\boxed{\text{ENTER}}$ $\boxed{\text{2nd}}$ $\boxed{\text{alpha}}$ $\boxed{\text{Y}}$ 1 $\boxed{\text{ENTER}}$.

```
fMax(y1,x,5,15)
              9.32332069352
Ans→x
              9.32332069352
y1
              54.6079078081

┌ fMin ┬ fMax ┬ arc ┬      ┬      ┐
```

We also observe that a relative minimum occurs in the interval $[-5, 5]$. Again, there are other intervals we could choose. To find the relative minimum in this interval, first press $\boxed{\text{2nd}}$ $\boxed{\text{CALC}}$ $\boxed{\text{MORE}}$ $\boxed{\text{F1}}$ to select the fMin feature from the CALC menu. (If the relative minimum is found immediately after the relative maximum was found above, the CALC menu will still be displayed and fMin can be selected by simply pressing $\boxed{\text{F1}}$.) Then enter the name of the function, the variable, and the endpoints of the interval. Press $\boxed{\text{2nd}}$ $\boxed{\text{alpha}}$ $\boxed{\text{Y}}$ 1 $\boxed{,}$ $\boxed{\text{x-VAR}}$ $\boxed{,}$ $\boxed{(-)}$ $\boxed{5}$ $\boxed{,}$ $\boxed{5}$ $\boxed{)}$ $\boxed{\text{ENTER}}$. We see that a relative minimum function value occurs when $x \approx 1.010010343$. To find the relative minimum value of the function press $\boxed{\text{STO▷}}$ $\boxed{\text{x-VAR}}$ $\boxed{\text{ENTER}}$ $\boxed{\text{2nd}}$ $\boxed{\text{alpha}}$ $\boxed{\text{Y}}$ 1 $\boxed{\text{ENTER}}$.

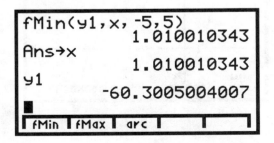

```
fMin(y1,x,-5,5)
              1.010010343
Ans→x
              1.010010343
y1
             -60.3005004007
■
┌ fMin ┬ fMax ┬ arc ┬      ┬      ┐
```

Chapter 4
Exponential and Logarithmic Functions

SOLVING EXPONENTIAL EQUATIONS

Section 4.2, page 307 Solve $e^t = 40$ using a grapher.

The Intersect feature is discussed on page 49 of this manual, and the ROOT feature is discussed on page 50.

EXPONENTIAL REGRESSION

Section 4.3, page 323 To find an exponential equation that models a set of data, enter the data in lists as described on page 53 of this manual. Then select EXPR (exponential regression) from the STAT CALC menu and find the regression equation by pressing STAT F1 ENTER ENTER F4 .

The grapher will return the values of a and b for an equation of the form $y = a \cdot b^x$. The function found can be evaluated for various values of x as described on page 46 of this manual.

Chapter 5
Integration

THE fnInt FEATURE

Definite integrals can be evaluated using the fnInt feature from the CALC menu.

Section 5.2, page 384 Evaluate $\int_{-1}^{2} (x^3 - 3x + 1)dx$ using the fnInt feature of a grapher.

First select the fnInt feature. From the home screen press $\boxed{\text{2nd}}$ $\boxed{\text{CALC}}$ $\boxed{\text{F5}}$ Then enter the function, the variable, and the lower and upper limits of integration. Press $\boxed{\text{x-VAR}}$ $\boxed{\wedge}$ $\boxed{3}$ $\boxed{-}$ $\boxed{3}$ $\boxed{\text{x-VAR}}$ $\boxed{+}$ $\boxed{1}$ $\boxed{,}$ $\boxed{\text{x-VAR}}$ $\boxed{,}$ $\boxed{(-)}$ $\boxed{1}$ $\boxed{,}$ $\boxed{2}$ $\boxed{)}$ $\boxed{\text{ENTER}}$. We find that $\int_{-1}^{2} (x^3 - 3x + 1)dx = 2.25$.

```
fnInt(x^3-3 x+1,x,-1,
2)
                        2.25
■
```

If the function has been entered on the $y(x) =$ screen, as y1 for instance, we can evaluate it by entering fnInt(y1, x, −1, 2) on the home screen. Press $\boxed{\text{2nd}}$ $\boxed{\text{CALC}}$ $\boxed{\text{F5}}$ $\boxed{\text{2nd}}$ $\boxed{\text{alpha}}$ $\boxed{\text{Y}}$ $\boxed{1}$ $\boxed{,}$ $\boxed{\text{x-VAR}}$ $\boxed{,}$ $\boxed{(-)}$ $\boxed{1}$ $\boxed{,}$ $\boxed{2}$ $\boxed{)}$ $\boxed{\text{ENTER}}$.

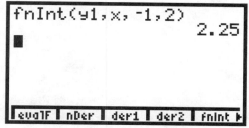

THE $\int f(x)dx$ FEATURE

Definite integrals can also be evaluated using the $\int f(x)$ feature from the CALC menu.

Section 5.2, page 384 Evaluate $\int_{-1}^{2} (x^3 - 3x + 1)dx$ using the $\int f(x)dx$ feature of a grapher.

First graph $y_1 = x^3 - 3x + 1$ in a rectangle that contains the interval $[-1, 2]$. We will use $[-3, 3, -6, 6]$. Then select the $\int f(x)$ feature from the GRAPH MATH menu by pressing $\boxed{\text{MORE}}$ $\boxed{\text{F1}}$ $\boxed{\text{F5}}$. Move the cursor to the lower limit of integration and press ENTER . Then move the cursor to the upper limit of integration and press ENTER again. The grapher returns the value of the definite integral on this interval.

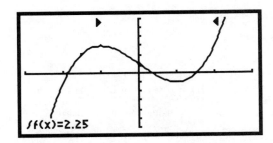

Chapter 6
Applications of Integration

STATISTICS

Section 6.8, page 475 The weights w of students in a calculus class are normally distributed with mean 150 lb and standard deviation 25 lb. Find the probability that a student's weight is from 160 lb to 180 lb.

The TI-85 does not have the capability to graph a probability density function. However, the program "dstrbutn.85p" in the stat folder found in the math category of the TI-85 Program Archive on the Texas Instruments web site, www.ti.com, provides a means for performing calculations involving the standard normal distribution. The program can be downloaded to your grapher.

Chapter 7
Functions of Several Variables

PARTIAL DERIVATIVES

Section 7.2, page 505 Given the function $f(x, y) = 3x^3y + 2xy$, use a grapher that finds derivatives of functions of one variable to find $f_x(-4, 1)$ and $f_y(2, 6)$.

To find $f_x(-4, 1)$, first find $f(x, 1)$:
$$f(x, y) = 3x^3y + 2xy$$
$$f(x, 1) = 3x^3 \cdot 1 + 2x \cdot 1$$
$$= 3x^3 + 2x$$

Now we have a function of one variable, so we use the nDer operation or the dy/dx operation to find the value of the derivative of this function when $x = -4$. (See pages 59 and 61 of this manual for the procedures to follow.)

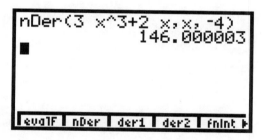

The procedures the grapher uses to calculate the derivative might not yield an exact answer. Note that the exact answer is 146, but the nDer operation yields 146.000003.

To find $f_y(2, 6)$, first find $f(2, y)$:
$$f(x, y) = 3x^3y + 2xy$$
$$f(2, y) = 3 \cdot 2^3y + 2 \cdot 2 \cdot y$$
$$= 24y + 4y = 28y$$

Now find the derivative of $f(y) = 28y$ when $y = 6$ using the nDer operation. Press 2nd CALC F2 2 8 2nd alpha Y , 2nd alpha Y , 6) ENTER.

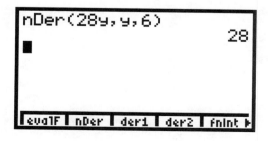

The TI-86
Graphics Calculator

Chapter 1
Functions, Graphs, and Models

GETTING STARTED

Press ON to turn on the TI-86 graphing calculator. (ON is the key at the bottom left-hand corner of the keypad.) You should see a blinking rectangle, or cursor, on the screen. If you do not see the cursor, try adjusting the display contrast. To do this, first press 2nd . (2nd is the yellow key in the left column of the keypad.) Then press and hold △ to increase the contrast or ▽ to decrease the contrast.

To turn the grapher off, press 2nd OFF . (OFF is the second operation associated with the ON key.) The grapher will turn itself off automatically after about five minutes without any activity.

Press 2nd MODE to display the MODE settings. (MODE is the second operation associated with the MORE key.) Initially you should select the settings on the left side of the display.

To change a setting on the Mode screen use ▽ or △ to move the cursor to the line of that setting. Then use ▷ or ◁ to move the blinking cursor to the desired setting and press ENTER . Press EXIT , CLEAR , or 2nd QUIT to leave the MODE screen. (QUIT is the second operation associated with the EXIT key.) Pressing EXIT , CLEAR , or 2nd QUIT will take you to the home screen where computations are performed.

It will be helpful to read the Quick Start section and Chapter 1: Operating the TI-86 in your TI-86 Guidebook before proceeding.

USING A MENU

A menu is a list of options that appears when a key is pressed. Thus, multiple options, and sometimes multiple menus, may be accessed by pressing one key. For example, the following screen appears when 2nd MATH is pressed. (MATH is the second operation associated with the × multiplication key.) We see several submenus at the bottom of the screen. The F1 - F5 keys at the top of the keypad are used to select options from this menu. The arrow to the right of MISC indicates that there are more choices. They can be seen by pressing MORE .

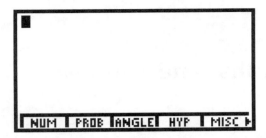

To choose the NUM submenu from the MATH menu press F1 . (If you pressed MORE to see the additional items on the MATH menu as described above, now press MORE again to see the first five items on the menu. Then press F1 to choose NUM.) When NUM is chosen, the original submenus move up on the screen and the items on the NUM submenu appear at the bottom of the screen.

When two rows of options are displayed like this, the top row is accessed by pressing 2nd followed by one of the keys F1 - F5 . These keystrokes access the second operations M1 - M5 associated with the F1 - F5 keys. The options on the bottom row are accessed by pressing one of the keys F1 - F5 . Absolute value, denoted "abs," is selected from the NUM submenu and copied to the home screen, for instance, by pressing F5 .

A menu can be removed from the screen by pressing EXIT . If both a menu and a submenu are displayed, press EXIT once to remove the submenu and twice to remove both.

SETTING THE VIEWING WINDOW

Section 1.1, page 7 (Page numbers refer to pages in the textbook.)

The viewing window is the portion of the coordinate plane that appears on the grapher's screen. It is defined by the minimum and maximum values of x and y: xMin, xMax, yMin, and yMax. The notation [xMin, xMax, yMin, yMax] is used in the text to represent these window settings or dimensions. For example, $[-12,\ 12,\ -8,\ 8]$ denotes a window that displays the portion of the x-axis from -12 to 12 and the portion of the y-axis from -8 to 8. In addition, the distance between tick marks on the axes is defined by the settings xScl and yScl. In this manual xScl and yScl will be assumed to be 1 unless noted otherwise. The setting xRes sets the pixel resolution. We usually select xRes $= 1$. The window corresponding to the settings $[-20,\ 30,\ -12,\ 20]$, xScl $= 5$, yScl $= 2$, xRes $= 1$, is shown below.

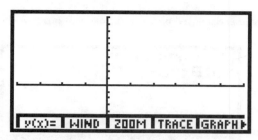

Press $\boxed{\text{GRAPH}}$ $\boxed{\text{F2}}$ to display the current window settings on your grapher. The standard settings $[-10, 10, -10, 10]$, xScl = 1, yScl = 1, are shown below.

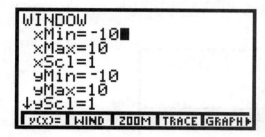 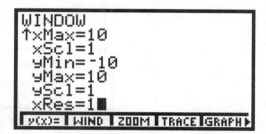

To change a setting, position the cursor beside the setting you wish to change and enter the new value. For example, to change from the standard settings to $[-20, 30, -12, 20]$, xScl = 5, yScl = 2, use the $\boxed{\triangle}$ and $\boxed{\triangledown}$ keys if necessary to position the cursor beside "xMax =" on the WINDOW screen. Then press $\boxed{(-)}$ 2 0 $\boxed{\text{ENTER}}$ 3 0 $\boxed{\text{ENTER}}$ 5 $\boxed{\text{ENTER}}$ $\boxed{(-)}$ 1 2 $\boxed{\text{ENTER}}$ 2 0 $\boxed{\text{ENTER}}$ 2 $\boxed{\text{ENTER}}$. You must use the $\boxed{(-)}$ key on the bottom of the keypad rather than the $\boxed{-}$ key in the right-hand column to enter a negative number. $\boxed{(-)}$ represents "the opposite of" or "the additive inverse of" whereas $\boxed{-}$ is the key for the subtraction operation. The $\boxed{\triangledown}$ key may be used instead of $\boxed{\text{ENTER}}$ after typing each window setting. To see the window shown above, press the $\boxed{\text{F5}}$ key on the top row of the keypad.

QUICK TIP: To return quickly to the standard window setting $[-10, 10, -10, 10]$, xScl = 1, yScl = 1, press $\boxed{\text{GRAPH}}$ $\boxed{\text{F3}}$ $\boxed{\text{F4}}$.

GRAPHS

After entering an equation and setting a viewing window, you can view the graph of an equation.

Section 1.1, page 8 Graph $y = x^3 - 5x + 1$ using a graphing calculator.

Equations are entered on the y(x) =, or equation-editor, screen. Press $\boxed{\text{GRAPH}}$ $\boxed{\text{F1}}$ to access this screen. If any of Plot 1, Plot 2, and Plot 3 is turned on (highlighted), turn it off by using the arrow keys to move the blinking cursor over the plot name and pressing $\boxed{\text{ENTER}}$. If there is currently an expression displayed for y1(x), clear it by positioning the cursor beside "y1(x) =" and pressing $\boxed{\text{CLEAR}}$. Do the same for expressions that appear on all other lines by using $\boxed{\triangledown}$ to move to a line and then pressing $\boxed{\text{CLEAR}}$. Then use $\boxed{\triangle}$ or $\boxed{\triangledown}$ to move the cursor beside "y1(x) =." Now press $\boxed{\text{F1}}$ $\boxed{\wedge}$ 3 $\boxed{-}$ 5 $\boxed{\text{F1}}$ $\boxed{+}$ 1 to enter the right-hand side of the equation in the y(x) = screen. The keystroke $\boxed{\text{F1}}$ can be replaced with $\boxed{\text{x-VAR}}$. When the y(x) = screen is displayed either keystroke will produce the variable x in an equation.

The standard $[-10, 10, -10, 10]$ window is a good choice for this graph. Either enter these dimensions in the WINDOW screen and then press $\boxed{\text{2nd}}$ $\boxed{\text{F5}}$ to see the graph or simply press $\boxed{\text{GRAPH}}$ $\boxed{\text{F3}}$ $\boxed{\text{F4}}$ to select the standard window and see the graph.

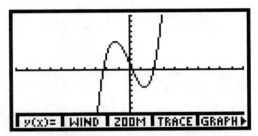

You can edit your entry if necessary. If, for instance, you pressed 6 instead of 5, use the $\boxed{\triangleleft}$ key to move the cursor to 6 and then press 5 to overwrite it. If you forgot to type the plus sign, move the cursor to 1, then press $\boxed{\text{2nd}}$ $\boxed{\text{INS}}$ $\boxed{+}$ to insert the plus sign before the 1. (INS is the second operation associated with the $\boxed{\text{DEL}}$ key.) You can continue to insert symbols immediately after the first insertion without pressing $\boxed{\text{2nd}}$ $\boxed{\text{INS}}$ again. If you typed 52 instead of 5, move the cursor to 2 and press $\boxed{\text{DEL}}$ to delete the 2.

An equation must be solved for y before it can be graphed on the TI-86.

Section 1.1, page 8 To graph $3x + 5y = 10$, first solve for y, obtaining $y = \dfrac{-3x + 10}{5}$. Then press $\boxed{\text{GRAPH}}$ $\boxed{\text{F1}}$ and clear any expressions that currently appear. Position the cursor beside "y1(x) =." Now press $\boxed{(}$ $\boxed{(-)}$ 3 $\boxed{\text{F1}}$ $\boxed{+}$ 1 0 $\boxed{)}$ $\boxed{\div}$ 5 to enter the right-hand side of the equation. Note that without the parentheses the expression $-3x + \dfrac{10}{5}$ would have been entered. As before, $\boxed{\text{F1}}$ can be replaced with $\boxed{\text{x-VAR}}$ to produce the variable x in the equation.

Select a viewing window and then press $\boxed{\text{F5}}$ to display the graph. You may change the viewing window as desired to reveal more or less of the graph. The standard window is shown here.

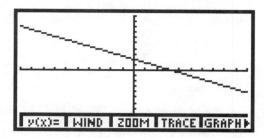

To graph $x = y^2$, first solve the equation for y : $y = \pm\sqrt{x}$. To obtain the entire graph of $x = y^2$, you must graph $y_1 = \sqrt{x}$ and $y_2 = -\sqrt{x}$ on the same screen. Press $\boxed{\text{GRAPH}}$ $\boxed{\text{F1}}$ and clear any expressions that currently appear. With the cursor beside "y1(x) =" press $\boxed{\text{2nd}}$ $\boxed{\sqrt{}}$ $\boxed{\text{F1}}$ or $\boxed{\text{2nd}}$ $\boxed{\sqrt{}}$ $\boxed{\text{x-VAR}}$. ($\boxed{\sqrt{}}$ is the second operation associated with the $\boxed{x^2}$ key.)

Now use $\boxed{\triangledown}$ to move the cursor beside "y2(x) =." We will describe two ways to enter $y_2 = -\sqrt{x}$. One is to enter the expression $-\sqrt{x}$ directly by pressing $\boxed{(-)}$ $\boxed{\text{2nd}}$ $\boxed{\sqrt{}}$ $\boxed{\text{F1}}$ or $\boxed{(-)}$ $\boxed{\text{2nd}}$ $\boxed{\sqrt{}}$ $\boxed{\text{x-VAR}}$.

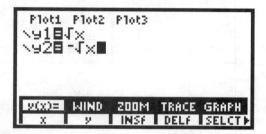

The other method of entering y_2 is based on the observation that $-\sqrt{x}$ is the opposite of the expression for y_1. That is, $y_2 = -y_1$. To enter this press $\boxed{(-)}$ $\boxed{\text{F2}}$ 1.

Select a viewing window and press $\boxed{\text{F5}}$ to display the graph. The window shown here is $[-2, 10, -5, 5]$.

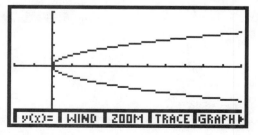

The top half is the graph of y_1, the bottom half is the graph of y_2, and together they yield the graph of $x = y^2$.

THE TABLE FEATURE

For an equation entered in the equation-editor screen, a table of x-and y-values can be displayed.

Section 1.2, page 18 Create a table of ordered pairs for the function $f(x) = x^3 - 5x + 1$.

Enter the function as $y_1 = x^3 - 5x + 1$ as described on page 79 of this manual. Once the equation in entered, press $\boxed{\text{TABLE}}$ $\boxed{\text{F2}}$ to display the table set-up screen. A minimum value of x can be chosen along with an increment for the x-value. To select a minimum x-value of 0.3 and an increment of 1, press $\boxed{\,.\,}$ 3 $\boxed{\triangledown}$ 1. The "Indpnt" setting should be "Auto." If it is not, use the $\boxed{\triangledown}$ key to position the blinking cursor over "Auto" on that line and then press $\boxed{\text{ENTER}}$. To display the table press $\boxed{\text{F1}}$.

   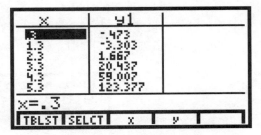

Use the $\boxed{\triangledown}$ and $\boxed{\triangle}$ keys to scroll through the table. For example, by using $\boxed{\triangledown}$ to scroll down we can see that $y_1 = 758.857$ when $x = 9.3$. Using $\boxed{\triangle}$ to scroll up, observe that $y_1 = -31.153$ when $x = -3.7$.

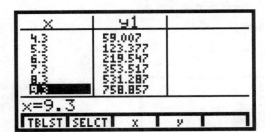   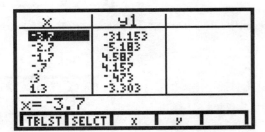

GRAPHS AND FUNCTION VALUES

Section 1.2, page 21 There are several ways to evaluate a function using a grapher. Three of them are described here. Given the function $f(x) = 2x^2 + x$, we will find $f(-2)$. First press $\boxed{\text{GRAPH}}$ $\boxed{\text{F1}}$ and enter the function as $y_1 = 2x^2 + x$. Now we will find $f(-2)$ using the TABLE feature. Press $\boxed{\text{TABLE}}$ $\boxed{\text{F2}}$ and select ASK mode by moving the cursor to "Indpnt: Ask" and pressing $\boxed{\text{ENTER}}$. In ASK mode, you supply the x-values and the grapher returns the corresponding y-values. The settings for TblStart nd ΔTbl are irrelevant in this mode. Press $\boxed{\text{F1}}$ and find $f(-2)$ by pressing $\boxed{(-)}$ 2 $\boxed{\text{ENTER}}$. We see that $y_1 = 6$ when $x = -2$, so $f(-2) = 6$.

 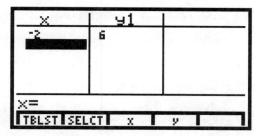

We can also use the EVAL feature from the GRAPH menu to find $f(-2)$. To do this, graph $y_1 = 2x^2 + x$ in a window that includes the x-value -2. We will use the standard window. Then press $\boxed{\text{MORE}}$ $\boxed{\text{MORE}}$ $\boxed{\text{F1}}$ to select the EVAL feature. Now supply the desired x-value by pressing $\boxed{(-)}$ 2. Press $\boxed{\text{ENTER}}$ to see x = -2, y = 6 at the bottom of the screen. Thus $f(-2) = 6$.

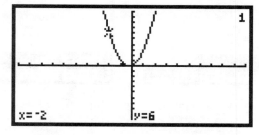

A third method for finding $f(-2)$ uses function notation directly. With $y_1 = 2x^2 + x$ entered on the y(x) = screen, go to the home screen by pressing $\boxed{\text{2nd}}$ $\boxed{\text{QUIT}}$. (QUIT is the second operation associated with the $\boxed{\text{EXIT}}$ key.) Now enter y1(-2) by pressing $\boxed{\text{2nd}}$ $\boxed{\text{alpha}}$ $\boxed{\text{Y}}$ 1 $\boxed{(}$ $\boxed{(-)}$ 2 $\boxed{)}$ $\boxed{\text{ENTER}}$. (alpha is the second operation associated with the blue ALPHA key, and Y is the alpha operation associated with the 0 numeric key.) Again we see that $y_1(-2) = 6$, or $f(-2) = 6$.

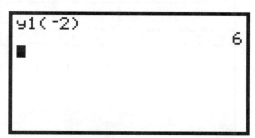

GRAPHING FUNCTIONS DEFINED PIECEWISE

Section 1.2, Example 9, page 22   Graph: $f(x) = \begin{cases} 4 \text{ for } x \leq 0, \\ 4 - x^2 \text{ for } 0 < x \leq 2, \\ 2x - 6 \text{ for } x > 2. \end{cases}$

We will enter the function using inequality symbols from the TEST menu. Press $\boxed{\text{GRAPH}}$ $\boxed{\text{F1}}$ to go to the equation-editor screen. Clear any entries that are present. Then position the cursor beside y1(x) = and press $\boxed{(}$ 4 $\boxed{)}$ $\boxed{(}$ $\boxed{\text{x-VAR}}$ $\boxed{\text{2nd}}$ $\boxed{\text{TEST}}$ $\boxed{\text{F4}}$ 0 $\boxed{)}$ + $\boxed{(}$ 4 $-$ $\boxed{\text{x-VAR}}$ $\boxed{x^2}$ $\boxed{)}$ $\boxed{(}$ 0 $\boxed{\text{F2}}$ $\boxed{\text{x-VAR}}$ $\boxed{\text{2nd}}$ $\boxed{\text{BASE}}$ $\boxed{\text{F4}}$ $\boxed{\text{F1}}$ $\boxed{\text{x-VAR}}$ $\boxed{\text{2nd}}$ $\boxed{\text{TEST}}$ $\boxed{\text{F2}}$ 2 $\boxed{)}$ + $\boxed{(}$ 2 $\boxed{\text{x-VAR}}$ $-$ 6 $\boxed{)}$ $\boxed{(}$ $\boxed{\text{x-VAR}}$ $\boxed{\text{F3}}$ 2 $\boxed{)}$. (TEST is the second operation associated with the 2 numeric key, and BASE is the second operation associated with the 1 numeric key.) The keystrokes $\boxed{\text{2nd}}$ $\boxed{\text{TEST}}$ $\boxed{\text{F4}}$ open the TEST menu and select item F4, the $\leq$ symbol, from that menu. Similarly, $\boxed{\text{2nd}}$ $\boxed{\text{TEST}}$ $\boxed{\text{F2}}$ opens the TEST menu and selects the symbol $<$. With the TEST menu displayed, we selected the $>$ symbol by pressing $\boxed{\text{F3}}$. The keystrokes $\boxed{\text{2nd}}$ $\boxed{\text{BASE}}$ $\boxed{\text{F4}}$ $\boxed{\text{F1}}$ selected "and" from the Boolean submenu ("Bool") of the BASE menu.

Now select DrawDot mode. If this is not done, a vertical line that is not part of the graph will appear. DrawDot mode can be selected in two ways. One is to press $\boxed{\text{MODE}}$, move the cursor over DrawDot, and press $\boxed{\text{ENTER}}$. DrawDot mode can also be selected on the y(x) = screen by pressing $\boxed{\text{GRAPH}}$ $\boxed{\text{F1}}$ $\boxed{\text{MORE}}$ $\boxed{\text{F3}}$ $\boxed{\text{F3}}$ $\boxed{\text{F3}}$ $\boxed{\text{F3}}$ $\boxed{\text{F3}}$ $\boxed{\text{F3}}$. Now the dot graph style icon appears to the left of y1.

 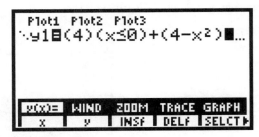

Choose and enter window dimensions and then press $\boxed{\text{F5}}$ to see the graph of the function. It is shown here in the window $[-5, 5, -3, 6]$.

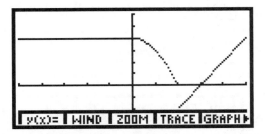

To return to connected, or DrawLine, mode press $\boxed{\text{GRAPH}}$ $\boxed{\text{MORE}}$ $\boxed{\text{F3}}$, highlight DrawLine, and press $\boxed{\text{ENTER}}$.

THE TRACE FEATURE

The TRACE feature can be used to display the coordinates of points on a graph.

Section 1.2, page 25 If you selected DrawDot mode for Example 9 above, return to Drawline mode now. Then enter the function $f(x) = x^3 - 5x + 1$ (see page 79 of this manual) and graph it in the window $[-5, 5, -10, 10]$. Now press $\boxed{\text{F4}}$. A blinking cursor appears on the graph and the coordinates of the point at which it is positioned are displayed at the bottom of the screen. Use the $\boxed{\triangleleft}$ and $\boxed{\triangleright}$ keys to move the cursor along the graph to see the coordinates of other points.

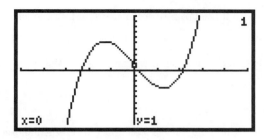 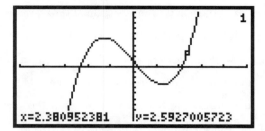

SQUARING THE VIEWING WINDOW

Section 1.4, page 41 In the standard window, the distance between tick marks on the y-axis is about 3/5 the distance between tick marks on the x-axis. It is often desirable to choose window dimensions for which these distances are the same, creating a "square" window. Any window in which the ratio of the length of the y-axis to the length of the x-axis is 3/5 will produce this effect. This can be accomplished by selecting dimensions for which $\text{yMax} - \text{yMin} = \frac{3}{5}(\text{xMax} - \text{xMin})$.

The standard window is shown on the left below and the square window $[-10, 10, -6, 6]$ is shown on the right. Observe that the distance between tick marks appears to be the same on both axes in the square window.

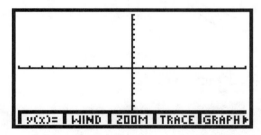 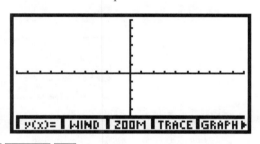

The window can also be squared by pressing $\boxed{\text{GRAPH}}$ $\boxed{\text{F3}}$ $\boxed{\text{MORE}}$ $\boxed{\text{F2}}$ to select the ZSQR feature from the GRAPH ZOOM menu. Starting with the standard window and selecting ZSQR produces the dimensions and the window shown below.

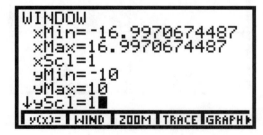 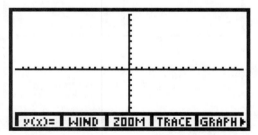

THE INTERSECT FEATURE

We can use the Intersect feature from the GRAPH MATH menu to solve equations.

Section 1.5, page 61 Solve the equation $x^3 = 3x + 1$ using the Intersect feature.

On the equation editor screen, clear any existing entries and then enter $\text{y1}(x) = x^3$ and $\text{y2}(x) = 3x + 1$. Now graph these equations in an appropriate window. One good choice is $[-3, 3, -10, 10]$. The solutions of the equation $x^3 = 3x + 1$ are the first coordinates of the points of intersection of these graphs. We will use the Intersect feature to find the leftmost point of intersection first. From the GRAPH screen, press $\boxed{\text{MORE}}$ $\boxed{\text{F1}}$ $\boxed{\text{MORE}}$ $\boxed{\text{F3}}$ to select ISECT from the GRAPH MATH menu. The query "First curve?" appears at the bottom of the screen. The blinking cursor is positioned on the graph of y1. This is indicated by the notation 1 in the upper right-hand corner of the screen. Press $\boxed{\text{ENTER}}$ to indicate that this is the first curve involved in the intersection. Next the query "Second curve?" appears at the bottom of the screen. The blinking cursor is now positioned on the graph of y2 and the notation 2 should appear in the top right-hand corner of the screen. Press $\boxed{\text{ENTER}}$ to indicate that this is the second curve. We identify the curves for the grapher since we could have as many as ten graphs on the screen at once. After we identify the second curve, the query "Guess?" appears at the bottom of the screen. Use the right and left arrow keys to move the blinking cursor close to the point of intersection of the graphs. This tells the grapher which point of intersection we are trying to find. When the cursor is positioned, press $\boxed{\text{ENTER}}$ a third time. Now the coordinates of the point of intersection appear at the bottom of the screen.

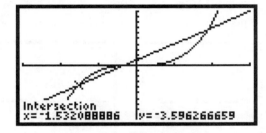

We see that, at the leftmost point of intersection, $x \approx -1.53$, so one solution of the equation is approximately -1.53. Repeat this process two times to find the coordinates of the other two points of intersection. We find that the other two solutions of the equation are approximately -0.35 and 1.88.

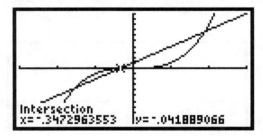
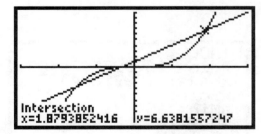

THE ROOT FEATURE

When an equation is expressed in the form $f(x) = 0$, it can be solved using the Root feature from the GRAPH MATH menu.

Section 1.5, page 61 Solve the equation $x^3 = 3x + 1$ using the Root feature.

First subtract $3x$ and 1 on both sides of the equation to obtain an equivalent equation with 0 on one side. We have $x^3 - 3x - 1 = 0$. The solutions of the equation $x^3 = 3x + 1$ are the values of x for which the function $f(x) = x^3 - 3x - 1$ is equal to 0. We can use the Root feature to find these values, or zeros.

On the equation-editor screen, clear any existing entries and then enter $y_1 = x^3 - 3x - 1$. Now graph the function in a viewing window that shows the x-intercepts clearly. One good choice is $[-3, 3, -5, 8]$. We see that the function has three zeros. They appear to be about -1.5, -0.5, and 2.

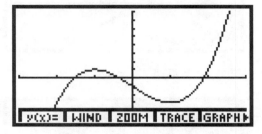

We will find the zero near -1.5 first. Press MORE F1 F1 to select the Root feature from the GRAPH MATH menu. We are prompted to select a left bound. This means that we must choose an x-value that is to the left of -1.5 on the x-axis. This can be done by using the left- and right-arrow keys to move the cursor to a point on the curve to the left of -1.5 or by keying in a value less than -1.5.

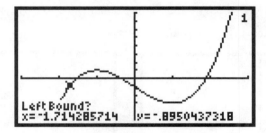

Once this is done press ENTER. Now we are prompted to select a right bound that is to the right of -1.5 on the x-axis. Again, this can be done by using the arrow keys to move the cursor to a point on the curve to the right of -1.5 or by keying in a value greater that -1.5.

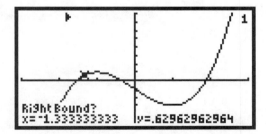

Press ENTER again. Finally we are prompted to make a guess as to the value of the zero. Move the cursor to a point close to the zero or key in a value.

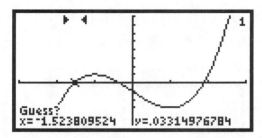

Press ENTER a third time. We see that $y = 0$ when $x \approx -1.53$, so -1.53 is a zero of the function.

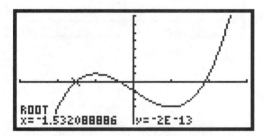

Select Root from the GRAPH MATH menu a second time to find the zero near -0.5 and a third time to find the zero near 2. We see that the other two zeros are approximately -0.35 and 1.88.

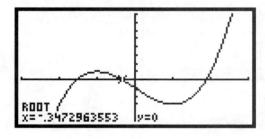   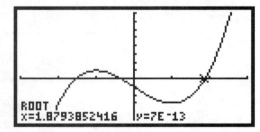

ABSOLUTE-VALUE FUNCTIONS

We can use the absolute-value option to perform computations involving absolute value and to graph absolute-value functions.

Section 1.5, page 65 Graph $f(x) = |x|$.

The absolute-value option is accessed from the MATH NUM (Number) menu. To enter $y = |x|$, first press GRAPH F1 to go to the equation-editor screen and then clear any existing entries. Now position the cursor beside y1 = and enter $|x|$ as abs x.

To do this, first press 2nd MATH F1 F5 to copy "abs" to the equation-editor screen. (MATH is the second operation associated with the $\times$ multiplication key.) Then press x-VAR. Choose an appropriate viewing window and graph the function. To use the standard window, press GRAPH F3 F4.

 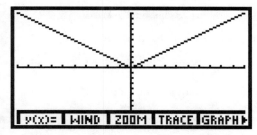

GRAPHING RADICAL FUNCTIONS

There are various way to enter radical expressions on the TI-86.

Section 1.5, page 67 We discussed entering an expression containing a square root on page 81 of this manual. If the radicand has more than one term, the entire radicand must be enclosed in parentheses. To enter $y_1 = \sqrt{x+2}$, for example, position the cursor beside y1 = on the equation editor screen and press $\boxed{\text{2nd}}$ $\boxed{\sqrt{}}$ $\boxed{(}$ $\boxed{\text{x-VAR}}$ $\boxed{+}$ 2 $\boxed{)}$.

Higher order radical expressions can be entered using the $\sqrt[x]{}$ option from the MATH MISC menu. We must enclose the radicand in parentheses if it contains more than one term. To enter $y_2 = \sqrt[3]{x-2}$, position the cursor beside y2 = on the equation-editor screen. Then press 3 to indicate that we are entering a cube root. Next press $\boxed{\text{2nd}}$ $\boxed{\text{MATH}}$ $\boxed{\text{F5}}$ $\boxed{\text{MORE}}$ $\boxed{\text{F4}}$ to select $\sqrt[x]{}$. Finally press $\boxed{(}$ $\boxed{\text{x-VAR}}$ $\boxed{-}$ 2 $\boxed{)}$ to enter the radicand.

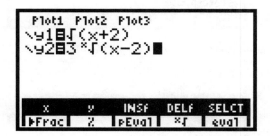

LINEAR REGRESSION

We can use the Linear Regression feature in the STAT CALC menu to fit a linear equation to a set of data.

Section 1.6, page 79 The following table lists data showing the price P of a one-day adult admission to Disney World for years since 1993.

Years, x (since 1993)	Price P of a One-Day Adult Admission to Disney World
0. 1993	\$34.00
1. 1994	\$36.00
2. 1995	\$37.00
3. 1996	\$40.81
4. 1997	\$42.14
5. 1998	\$44.52

(a) Fit a regression line to the data using the REGRESSION feature on a grapher.

(b) Graph the regression line with the scatterplot.

(c) Use the model to predict the price of a one-day adult admission in 2000 ($x = 7$).

(a) We will enter the data as ordered pairs on the STAT list editor screen. To go to this screen and clear any existing lists first press ⟦2nd⟧ ⟦STAT⟧ ⟦F2⟧. (STAT is the second operation associated with the ⟦+⟧ key.) Then use the arrow keys to move up to highlight "xStat" and press ⟦CLEAR⟧ ⟦ENTER⟧. Do the same for "yStat."

Once the lists are cleared, we can enter the data points. We will enter the number of years since 1993 in xStat and the prices in yStat. Position the cursor at the top of the xStat column, below the xStat heading. To enter 0 press 0 ⟦ENTER⟧. Continue typing the x-values 1 through 5, each followed by ⟦ENTER⟧. The entries can be followed by ⟦▽⟧ rather than ⟦ENTER⟧ if desired. Press ⟦▷⟧ to move to the top of the yStat column. Type the prices 34.00, 36.00, and so on in succession, each followed by ⟦ENTER⟧ or ⟦▽⟧. Note that the coordinates of each point must be in the same position in both lists.

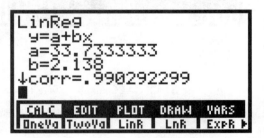

The grapher's linear regression feature can be used to fit a linear equation to the data. Once the data have been entered in the lists, go to the home screen and press ⟦2nd⟧ ⟦STAT⟧ ⟦F1⟧ ⟦F3⟧ ⟦ENTER⟧ to select LinReg($ax + b$) from the STAT CALC menu and to display the coefficients a and b of the regression equation $y = a + bx$.

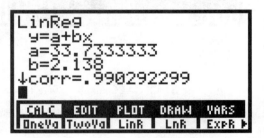

Note that we also see the coefficient of correlation, denoted "corr." This number indicates how well the regression line fits the data. The closer the absolute value of this number is to 1, the better the fit. If we press ⟦EXIT⟧ to clear the STAT CALC menu from the bottom of the screen, we also see "n = 6." This indicates that 6 data points were used to find the regression equation.

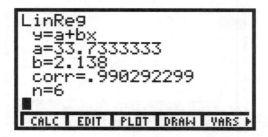

Immediately after the regression equation is found it can be copied to the equation-editor screen. We will copy it as y1. Note that any previous entry for y1 must have been cleared first. Press $\boxed{\text{GRAPH}}$ $\boxed{\text{F1}}$ and position the cursor beside y1 =. Then press $\boxed{\text{2nd}}$ $\boxed{\text{CATLG-VARS}}$ $\boxed{\text{MORE}}$ $\boxed{\text{MORE}}$ $\boxed{\text{F4}}$. (CATLG-VARS is the second operation associated with the $\boxed{\text{CUSTOM}}$ key.) Use the $\boxed{\bigtriangledown}$ key to position the selection cursor beside "RegEq" and press $\boxed{\text{ENTER}}$. These keystrokes paste the RegEq (regression equation) into the equation-editor screen.

Alternatively, before the regression equation is found, it is possible to select a y-variable to which it will be stored on the equation editor screen. After the data have been stored in the lists and the equation previously entered as y1 has been cleared, press $\boxed{\text{2nd}}$ $\boxed{\text{STAT}}$ $\boxed{\text{F1}}$ $\boxed{\text{F3}}$ $\boxed{\text{2nd}}$ $\boxed{\text{alpha}}$ $\boxed{\text{Y}}$ 1 $\boxed{\text{ENTER}}$. The coefficients of the regression equation will be displayed on the home screen, and the regression equation will also be stored as y1 on the equation-editor screen.

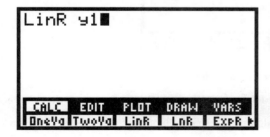 

(b) To plot the data points, we turn on the STAT PLOT feature. To access the STAT PLOT screen, press $\boxed{\text{2nd}}$ $\boxed{\text{STAT}}$ $\boxed{\text{F3}}$. Choose Plot 1 by pressing $\boxed{\text{F1}}$. Position the cursor over "On" and press $\boxed{\text{ENTER}}$. The entries Type, Xlist, and Ylist should be as shown below. To change an entry position the cursor beside it and make the appropriate choice from the menu that appears at the bottom of the screen. The last item, Mark, allows us to choose a box, a cross, or a dot for each point. Here we have selected a box.

The plot can also be turned on from the "y(x) =" screen. Press GRAPH F1 to go to this screen. Then, assuming that Plot 1 has not yet been turned on and that the desired settings are currently entered for Plot 1 on the STAT PLOT screen, position the cursor over Plot 1 and press ENTER . Plot 1 will now be highlighted.

Now select a viewing window. The years range from 0 through 5 and the prices range from \$34.00 through \$44.52, so one good choice is $[-1, 6, 30, 50]$. To see the plotted points and the regression line, press F5 .

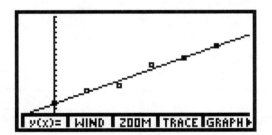

QUICK TIP: Instead of entering the window dimensions directly, we can press GRAPH F3 MORE F5 after entering the coordinates of the points in lists, turning on Plot 1, and selecting Type, Xlist, Ylist, and Mark. This activates the ZDATA operation which automatically defines a viewing window that displays all of the points and also displays the graph.

(c) To predict the price of a one-day adult admission in 2000, evaluate the regression equation for $x = 7$. (2000 is 7 years after 1993.) We will use the Forecast feature from the STAT menu. Press 2nd STAT MORE F1 . The Forecast screen will be displayed with the cursor beside x =. Press 7 ENTER . The cursor moves beside y =. Press F5 to select SOLVE. When $x = 7$, $y \approx 48.70$, so we predict that the price of a one-day adult admission to Disney World will be about \$48.70 in 2000.

We could also use any of the methods for evaluating a function presented earlier in this chapter. (See pages 82 and 83.)

POLYNOMIAL REGRESSION

The TI-86 has the capability to use regression to fit quadratic, cubic, and quartic functions to data.

Section 1.6, page 81 The following chart relates the number of live births to women of a particular age.

Age, x	Average Number of Live Births per 1000 Women
16	34
18.5	86.5
22	111.1
27	113.9
32	84.5
37	35.4
42	6.8

(a) Fit a quadratic function to the data using the REGRESSION feature on a grapher.

(b) Make a scatterplot of the data. Then graph the quadratic function with the scatterplot.

(c) Fit a cubic function to the data using the REGRESSION feature on a grapher.

(d) Make a scatterplot of the data. Then graph the cubic function with the scatterplot.

(e) Decide which function seems to fit the data better.

(f) Use the function from part (e) to estimate the average number of live births by women of ages 20 and 30.

(a) First enter the data with the ages in xStat and the average number of live births per 1000 women in yStat. (See page 90 of this manual.) Go to the home screen and select quadratic regression, denoted P2Reg, from the STAT CALC menu. Press | 2nd | | STAT | | F1 | | MORE | | F4 | | ENTER | to do this.

The grapher returns the coefficients of a quadratic function of the form $f(x) = ax^2 + bx + c$. Use the $\boxed{\triangleright}$ key to see the numbers that do not appear initially. Rounding the coefficients to two decimal places, we obtain the function $f(x) = -0.49x^2 + 25.95x - 238.49$.

(b) Graph the function along with the scatterplot as described on pages 91 and 92 of this manual.

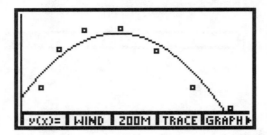

(c) Once the data are entered, fit a cubic function to it by pressing $\boxed{\text{2nd}}$ $\boxed{\text{STAT}}$ $\boxed{\text{F1}}$ $\boxed{\text{MORE}}$ $\boxed{\text{F5}}$ $\boxed{\text{ENTER}}$. These keystrokes select the cubic regression feature, denoted P3Reg, from the STAT CALC menu and display the coefficients of a cubic function of the form $f(x) = ax^3 + bx^2 + cx + d$.

Rounding the coefficients to two decimal places, we have $f(x) = 0.03x^3 - 3.22x^2 + 101.18x - 886.93$.

(d) Graph the function along with the scatterplot as described on pages 91 and 92 of this manual.

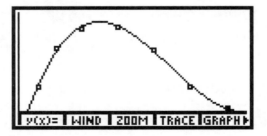

(e) The graph of the cubic function appears to fit the data better than the graph of the quadratic function.

(f) We can use any of the methods described earlier in this chapter to evaluate the function. We will use the Forecast feature from the STAT menu. (See page 92 of this manual.)

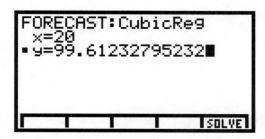 

We estimate that there will be about 100 live births per 1000 20-year-old women and about 97 live births per 1000 30-year-old women.

A quartic function can be fit to data by selecting quartic regression, denoted P4Reg, from the STAT CALC menu. The grapher returns the coefficients of a function $f(x) = ax^4 + bx^3 + cx^2 + dx + e$.

Chapter 2
Differentiation

ZOOMING IN

The ZIN (Zoom In) operation from the GRAPH ZOOM menu can be used to enlarge a portion of a graph.

Section 2.2, page 112 Verify graphically that $\lim\limits_{x \to 0} \dfrac{\sqrt{x+1}-1}{x} = 0.5$.

First graph $y = \dfrac{\sqrt{x+1}-1}{x}$ in a window that shows the portion of the graph near $x = 0$. One good choice is $[-2, 5, -1, 2]$. Be sure that the plots are turned off and that all other entries on the equation-editor screen are cleared. Note that the radicand, $x + 1$, must be enclosed in parentheses. That is, we enter the equation as $y = (\sqrt{(x+1)} - 1)/x$. Without the parentheses around the radicand the grapher would read the expression as $y = \dfrac{\sqrt{x}+1-1}{x}$. Then press $\boxed{\text{F4}}$ and use the $\boxed{\triangleleft}$ and $\boxed{\triangleright}$ keys to move the trace cursor to a point on the curve near $x = 0$.

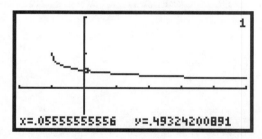

Now select the ZIN operation from the GRAPH ZOOM menu by pressing $\boxed{\text{GRAPH}}$ $\boxed{\text{F3}}$ $\boxed{\text{F2}}$ $\boxed{\text{ENTER}}$. This enlarges the portion of the graph near $x = 0$. We can now press $\boxed{\text{GRAPH}}$ $\boxed{\text{F4}}$ and use the $\boxed{\triangleleft}$ and $\boxed{\triangleright}$ keys to trace the curve near $x = 0$.

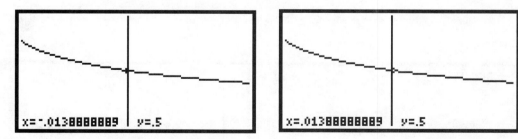

The Zoom In operation can be used as many times as desired in order to verify the result.

THE nDer OPERATION

The TI-86 can be used to find the slope of a line tangent to a curve at a specific point. That is, it can find the derivative of a function $f(x)$ for a specific value of x.

Section 2.4, page 134 For the function $f(x) = x(100 - x)$, find $f'(70)$.

We will use the nDer (numerical derivative) operation from the CALC menu. Select this operation by pressing 2nd CALC F2. Now enter the function, the variable, and the value at which the derivative is to be evaluated, all separated by commas. Press x-VAR (1 0 0 − x-VAR) , x-VAR , 7 0) ENTER. Note that the grapher supplies a left parenthesis after "nDer" and we close the parentheses with a right parenthesis after entering 70. We see that $f'(70) = -40$.

THE TanLn FEATURE

We can draw a line tangent to a curve at a given point using the Tangent feature from the DRAW submenu of the GRAPH menu.

Section 2.4, page 134 Draw the line tangent to the graph of $f(x) = x(100 - x)$ at $x = 70$.

First graph $y_1 = x(100 - x)$ in a window that shows the portion of the curve near $x = 70$. One good choice is $[-10, 100, -10, 3000]$, xScl = 10, yScl = 1000. Be sure to clear any functions that were previously entered and turn off the plots. Now select the TanLn feature from the GRAPH DRAW menu and instruct the grapher to draw the line tangent to the graph of y_1 at $x = 70$. To do this press MORE F2 MORE MORE MORE F2 2nd alpha Y 1 , 7 0) ENTER. Note that the grapher supplies a left parenthesis after "TanLn" and we close the parentheses with a right parenthesis after entering 70.

 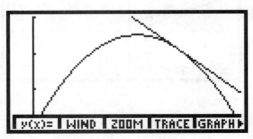

Use the ClrDraw (clear drawing) operation from the GRAPH DRAW menu to clear the graph of the tangent line from the GRAPH screen. Press GRAPH MORE F2 MORE MORE F1.

See page 99 of this manual for a procedure that draws a tangent line directly from the GRAPH screen.

THE dy/dx OPERATION

We can find the derivative of a function at a specific point directly from the GRAPH screen.

Section 2.5, page 139 For the function $f(x) = x\sqrt{4 - x^2}$, find dy/dx at a specific point.

First graph $y = x\sqrt{(4 - x^2)}$. We will use the window $[-3, 3, -4, 4]$. Then select the dy/dx operation from the GRAPH MATH menu by pressing $\boxed{\text{MORE}}$ $\boxed{\text{F1}}$ $\boxed{\text{F2}}$. We see the graph with a cursor positioned on it in the middle of the window which, in this case, is at $(0, 0)$.

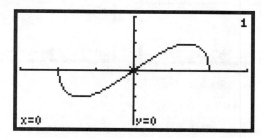

To find the value of dy/dx at a specific point either move the cursor to the desired point or key in its x-coordinate. For example, if we move the cursor to the point $(1.2380952381, 1.9446847395)$ and press $\boxed{\text{ENTER}}$ we find that $dy/dx = 0.5947897471$ at this point.

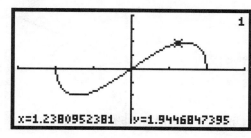 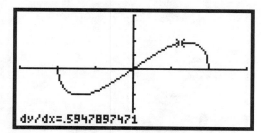

To find dy/dx for $x = 1$, select dy/dx from the GRAPH MATH menu by pressing $\boxed{\text{GRAPH}}$ $\boxed{\text{MORE}}$ $\boxed{\text{F1}}$ $\boxed{\text{F2}}$. Then press 1 $\boxed{\text{ENTER}}$. We see that $dy/dx = 1.1547005384$ at this point.

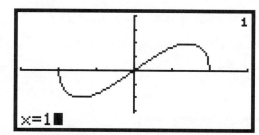 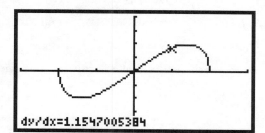

MORE ON TANGENT LINES

We can draw a line tangent to a curve at a specific point directly from the GRAPH screen.

Section 2.5, page 139 Draw the line tangent to the graph of $f(x) = x\sqrt{4 - x^2}$ at a specific point.

First graph $y = x\sqrt{(4 - x^2)}$ in an appropriate window such as $[-3, 3, -4, 4]$. Then press $\boxed{\text{F4}}$ and use the $\boxed{\triangleleft}$ and $\boxed{\triangleright}$

keys to move the cursor to the desired point. Now select the TANLN feature from the MATH submenu of the GRAPH menu and see the tangent line by pressing GRAPH MORE F1 MORE MORE F1 ENTER. The slope of the curve at the point of tangency is also displayed.

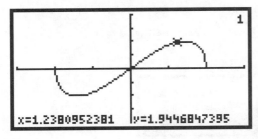 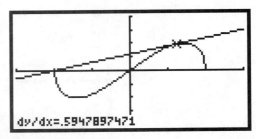

Use the ClrDraw (clear drawing) operation from the GRAPH DRAW menu to clear the graph of the tangent line from the GRAPH screen. Press GRAPH MORE F2 MORE MORE F1.

Rather than using the TRACE feature to move the cursor to a point of tangency, we can also enter the x-coordinate of the point directly. For example, to graph the line tangent to the curve at $x = 1$, from the GRAPH screen press MORE F1 MORE MORE F1 as before to select the TANLN feature. Then press 1 ENTER to enter the x-coordinate, 1, and see the tangent line and the slope of the curve at the point of tangency.

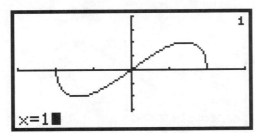 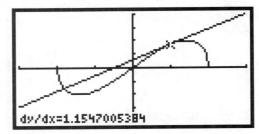

The graph of the tangent line can be cleared as described above.

ENTERING THE SUM OF TWO FUNCTIONS

Section 2.5, page 142 The Technology Connection on this page involves finding the derivative of a function $y_3 = y_1 + y_2$, where $y_1 = x(100 - x)$ and $y_2 = x\sqrt{100 - x^2}$. To enter y_3, first press GRAPH F1 to go to the equation-editor screen. Clear any entries present and turn off the plots. Then enter y_1 and y_2. To enter $y_3 = y_1 + y_2$, position the cursor beside "y3 =" and press F2 1 + F2 2.

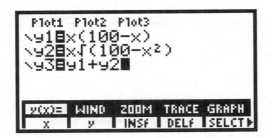

DESELECTING FUNCTIONS; GRAPH STYLES

Section 2.7, page 160 The Technology Connection on this page involves deselecting a function and also using different graph styles for two functions. First enter $y_1 = \dfrac{x^2 - 3x}{x - 1}$, $y_2 = \dfrac{x^2 - 2x + 3}{(x - 1)^2}$, and $y_3 = \text{nDer}(y1, x, x)$. Since we want to see only the graphs of y_2 and y_3, we will deselect y_1. To do this, position the cursor anywhere in the function y1 and press $\boxed{\text{2nd}}$ $\boxed{\text{F5}}$ to choose SELCT (Select). Note that the equals sign is no longer highlighted. This indicates that y_1 has been deselected and, thus, its graph will not appear with the graphs of y_2 and y_3, the functions which continue to be selected.

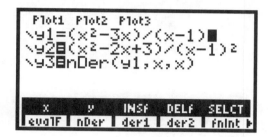

To select y_1 again, position the cursor anywhere is the function y1 and press $\boxed{\text{GRAPH}}$ $\boxed{\text{F1}}$ $\boxed{\text{F5}}$. The equals sign will be highlighted now, indicating that the function is selected.

With y_1 deselected, we can graph y_2 and y_3 using different graph styles to determine whether the graphs coincide. When the grapher is in LineDraw mode, equations are graphed with a solid line. We will keep the solid line graph style for the graph of y_2 and select the path style for the graph of y_3. After the graph of y_2 is drawn, a circular cursor will trace the leading edge of the graph of y_3 and draw its path. (This assumes that the grapher is set in Sequential mode.)

To select the path style for y_3, first position the cursor in the function y_3. Then press $\boxed{\text{MORE}}$ $\boxed{\text{F3}}$ $\boxed{\text{F3}}$ $\boxed{\text{F3}}$ $\boxed{\text{F3}}$. The path icon will appear beside y_3. The window below shows y_1 deselected, y_2 with the solid line graph style selected, and y_3 with the path style selected.

To see the graphs of y_2 and y_3 in the standard window, press $\boxed{\text{GRAPH}}$ $\boxed{\text{F3}}$ $\boxed{\text{F4}}$.

Chapter 3
Applications of Differentiation

THE FMAX AND FMIN FEATURES

Section 3.1, page 195   Use the FMAX and FMIN features of a grapher to approximate the relative extrema of $f(x) = -0.4x^3 + 6.2x^2 - 11.3x - 54.8$.

First graph $y_1 = -0.4x^3 + 6.2x^2 - 11.3x - 54.8$ in a window that displays the relative extrema of the function. Trial and error reveals that one good choice is $[-10, 20, -100, 150]$, xScl $= 5$, yScl $= 50$. Observe that a relative maximum occurs near $x = 10$ and a relative minimum occurs near $x = 1$.

To find the relative maximum, first press $\boxed{\text{MORE}}$ $\boxed{\text{F1}}$ $\boxed{\text{F5}}$ to select the FMAX feature from the GRAPH MATH menu. We are prompted to select a left bound for the relative maximum. This means that we must choose an x-value that is to the left of the x-value of the point where the relative maximum occurs. This can be done by using the left- and right-arrow keys to move the cursor to a point to the left of the relative maximum or by keying in an appropriate value.

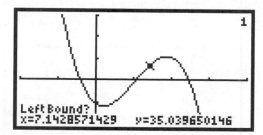

Once this is done, press $\boxed{\text{ENTER}}$. Now we are prompted to select a right bound. We move the cursor to a point to the right of the relative maximum or we key in an appropriate value.

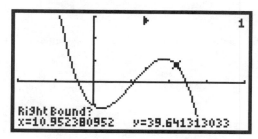

Press $\boxed{\text{ENTER}}$ again. Finally we are prompted to guess the x-value at which the relative maximum occurs. Move the cursor close to the relative maximum point or key in an x-value.

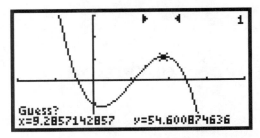

Press ENTER a third time. We see that a relative maximum function value of approximately 54.61 occurs when $x \approx 9.32$.

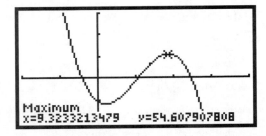

To find the relative minimum, select the FMIN feature from the GRAPH MATH menu by pressing GRAPH MORE F1 F4. Select left and right bounds for the relative minimum and guess the x-value at which it occurs as described above. We see that a relative minimum function value of approximately -60.30 occurs when $x \approx 1.01$.

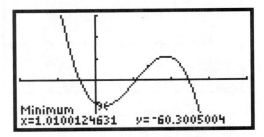

THE fMax AND fMin FEATURES

The fMax and fMin features can be used to calculate the x-values at which relative maximum and minimum values of a function occur over a specified closed interval.

Section 3.1, page 195 Use the fMax and fMin features of a grapher to approximate the relative extrema of $f(x) = -0.4x^3 + 6.2x^2 - 11.3x - 54.8$.

First enter the function as y_1 and graph it as described above. Observe that a relative maximum occurs in the interval [5, 15]. There are other intervals we could use. Keep in mind that the larger the interval, the longer it takes the grapher to return an x-value.

Now go to the home screen and press 2nd CALC MORE F2 to select the fMax feature from the CALC menu. Enter the name of the function, the variable, and the left and right endpoints of the interval on which the relative maximum occurs by pressing 2nd alpha Y 1 , x-VAR , 5 , 1 5) . Press ENTER to find that the relative maximum occurs when $x \approx 9.32332128289$. To find the relative maximum value of the function, we evaluate the function for this value of x. Press 2nd alpha Y 1 (2nd ANS) ENTER . (ANS is the second operation associated with the (−) key.) The keystrokes 2nd ANS cause the grapher to use the previous answer, 9.32332128289, as the value for x in y_1.

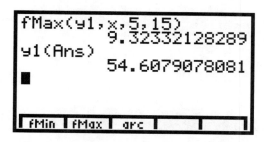

We also observe that a relative minimum occurs in the interval $[-5, 5]$. Again, there are other intervals we could choose. To find the relative minimum in this interval, first press 2nd CALC MORE F1 to select the fMin feature from the CALC menu. Then enter the name of the function, the variable, and the endpoints of the interval. Press 2nd alpha Y 1 , x-VAR , (−) 5 , 5) ENTER . We see that a relative minimum function value occurs when $x \approx 1.01001202891$. To find the relative minimum value of the function press 2nd alpha Y 1 (2nd ANS) ENTER .

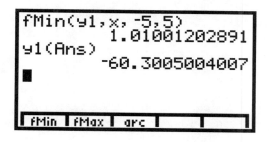

Chapter 4
Exponential and Logarithmic Functions

SOLVING EXPONENTIAL EQUATIONS

Section 4.2, page 307 Solve $e^t = 40$ using a grapher.

The Intersect feature is discussed on page 85 of this manual, and the Root feature is discussed on page 86.

EXPONENTIAL REGRESSION

Section 4.3, page 323 To find an exponential equation that models a set of data, enter the data in lists as described on page 90 of this manual. Then select ExpR (exponential regression) from the STAT CALC menu and find the regression equation by pressing 2nd STAT F1 F5 ENTER .

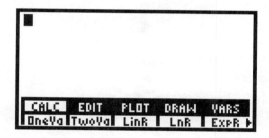

The grapher will return the values of a and b for an equation of the form $y = a \cdot b^x$. The function found can be evaluated for various values of x as described on pages 82, 83, and 92 of this manual.

Chapter 5
Integration

THE fnInt FEATURE

Definite integrals can be evaluated using the fnInt feature from the CALC menu.

Section 5.2, page 384 Evaluate $\int_{-1}^{2} (x^3 - 3x + 1)dx$ using the fnInt feature of a grapher.

First select the fnInt feature. Press $\boxed{\text{2nd}}$ $\boxed{\text{CALC}}$ $\boxed{\text{F5}}$. Then enter the function, the variable, and the lower and upper limits of integration. Press $\boxed{\text{x-VAR}}$ $\boxed{\wedge}$ $\boxed{3}$ $\boxed{-}$ $\boxed{3}$ $\boxed{\text{x-VAR}}$ $\boxed{+}$ $\boxed{1}$ $\boxed{,}$ $\boxed{\text{x-VAR}}$ $\boxed{,}$ $\boxed{(-)}$ $\boxed{1}$ $\boxed{,}$ $\boxed{2}$ $\boxed{)}$ $\boxed{\text{ENTER}}$. We find that $\int_{-1}^{2} (x^3 - 3x + 1)dx = 2.25$.

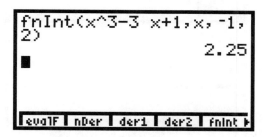

If the function has been entered on the y(x) = screen, as y_1 for instance, we can evaluate it by entering fnInt(y1, x, −1, 2) on the home screen. Press $\boxed{\text{2nd}}$ $\boxed{\text{CALC}}$ $\boxed{\text{F5}}$ $\boxed{\text{2nd}}$ $\boxed{\text{alpha}}$ $\boxed{\text{Y}}$ $\boxed{1}$ $\boxed{,}$ $\boxed{\text{x-VAR}}$ $\boxed{,}$ $\boxed{(-)}$ $\boxed{1}$ $\boxed{,}$ $\boxed{2}$ $\boxed{)}$ $\boxed{\text{ENTER}}$.

 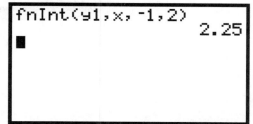

THE $\int f(x)$ FEATURE

Definite integrals can also be evaluated using the $\int f(x)$ feature from the GRAPH MATH menu.

Section 5.2, page 384 Evaluate $\int_{-1}^{2} (x^3 - 3x + 1)dx$ using the $\int f(x)$ feature of a grapher.

First graph $y_1 = x^3 - 3x + 1$ in a window that contains the interval $[-1, 2]$. We will use $[-3, 3, -6, 6]$. Then select the $\int f(x)$ feature from the GRAPH MATH menu by pressing $\boxed{\text{MORE}}$ $\boxed{\text{F1}}$ $\boxed{\text{F3}}$. We are prompted to enter the lower limit of integration. Press $\boxed{(-)}$ 1.

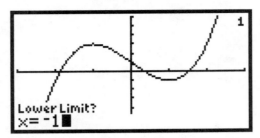

Then press ENTER . Now enter the upper limit of integration by pressing 2.

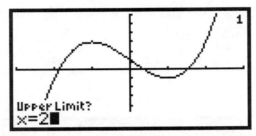

Press ENTER again. The grapher shades the area above and below the curve on $[-1, 2]$ and returns the value of the definite integral on this interval.

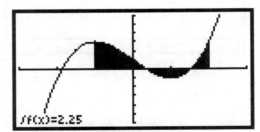

Chapter 6
Applications of Integration

STATISTICS

Section 6.8, page 475 The weights w of students in a calculus class are normally distributed with mean 150 lb and standard deviation 25 lb. Find the probability that a student's weight is from 160 lb to 180 lb.

The TI-86 does not have the capability to graph a probability density function. However, there are programs in the stat folder found in the math category of the TI-86 Program Archive on the Texas Instruments web site, www.ti.com, that provide a means for performing calculations involving the standard normal distribution. These programs can be downloaded to your grapher.

Chapter 7
Functions of Several Variables

PARTIAL DERIVATIVES

Section 7.2, page 505 Given the function $f(x, y) = 3x^3y + 2xy$, use a grapher that finds derivatives of functions of one variable to find $f_x(-4, 1)$ and $f_y(2, 6)$.

To find $f_x(-4, 1)$, first find $f(x, 1)$:
$$f(x, y) = 3x^3y + 2xy$$
$$f(x, 1) = 3x^3 \cdot 1 + 2x \cdot 1$$
$$= 3x^3 + 2x$$

Now we have a function of one variable, so we use the nDer operation or the dy/dx operation to find the value of the derivative of this function when $x = -4$. (See pages 98 and 99 of this manual for the procedures to follow.)

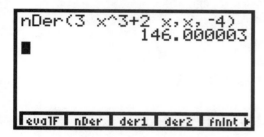
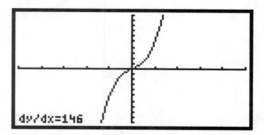

The procedures the grapher uses to calculate the derivative might not yield an exact answer. Note that the exact answer is 146, but the nDer operation yields 146.000003.

To find $f_y(2, 6)$, first find $f(2, y)$:
$$f(x, y) = 3x^3y + 2xy$$
$$f(2, y) = 3 \cdot 2^3y + 2 \cdot 2 \cdot y$$
$$= 24y + 4y = 28y$$

Now find the derivative of $f(y) = 28y$ when $y = 6$ using the nder operation. Press $\boxed{\text{2nd}}$ $\boxed{\text{CALC}}$ $\boxed{2}$ 2 8 $\boxed{\text{2nd}}$ $\boxed{\text{alpha}}$ $\boxed{\text{Y}}$ $\boxed{,}$ $\boxed{\text{2nd}}$ $\boxed{\text{alpha}}$ $\boxed{\text{Y}}$ $\boxed{,}$ $\boxed{6}$ $\boxed{)}$ $\boxed{\text{ENTER}}$.

We can also replace y with x and find the derivative of $f(x) = 28x$ when $x = 6$ using the dy/dx operation.

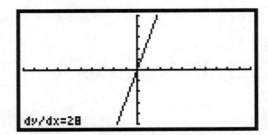

The TI-89
Graphics Calculator

Chapter 1
Functions, Graphs, and Models

GETTING STARTED

Press $\boxed{\text{ON}}$ to turn on the TI-89 graphing calculator. $\Big($ $\boxed{\text{ON}}$ is the key at the bottom left-hand corner of the keypad.) The home screen is displayed. You should see a row of boxes at the top of the screen and two horizontal lines with lettering below them at the bottom of the screen. If you do not see anything, try adjusting the display contrast. To do this, first press $\boxed{\diamond}$. $\Big($ $\boxed{\diamond}$ is the key in the left column of the keypad with a green diamond inside a green border. All operations associated with the $\boxed{\diamond}$ key are printed on the keyboard in green, the same color as the $\boxed{\diamond}$ key.) Then press $\boxed{+}$ to darken the display or $\boxed{-}$ to lighten the display. Be sure to use the black $\boxed{-}$ key in the right column of the keypad rather than the gray $\boxed{(-)}$ key on the bottom row.

One way to turn the grapher off is to press $\boxed{\text{2nd}}$ $\boxed{\text{OFF}}$. (OFF is the second operation associated with the $\boxed{\text{ON}}$ key. All operations accessed by using the $\boxed{\text{2nd}}$ key are printed on the keyboard in yellow, the same color as the $\boxed{\text{2nd}}$ key.) When you turn the TI-89 on again the home screen will be displayed regardless of the screen that was displayed when the grapher was turned off. $\boxed{\text{2nd}}$ $\boxed{\text{OFF}}$ cannot be used to turn off the grapher if an error message is displayed. The grapher can also be turned off by pressing $\boxed{\diamond}$ $\boxed{\text{OFF}}$. This will work even if an error message is displayed. When the TI-89 is turned on again the display will be exactly as it was when it was turned off. The grapher will turn itself off automatically after several minutes without any activity. When this happens the display will be just as you left it when you turn the grapher on again.

From top to bottom, the home screen consists of the tool bar, the large history area where entries and their corresponding results are displayed, the entry line where expressions or instructions are entered, and the status line which shows the current state of the calculator. These areas will be discussed in more detail as the need arises.

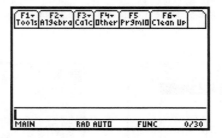

Press $\boxed{\text{MODE}}$ to display the MODE settings. Modes that are not currently valid, due to the existing choices of settings, are dimmed. Initially you should select the settings shown below.

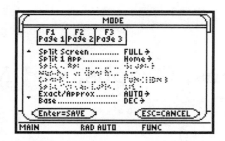

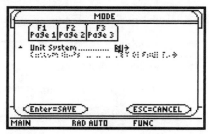

To change a setting on the Mode screen use ∇ or $\triangle$ to move the cursor to the line of that setting. Then use $\triangleright$ to display the options. Press the number of the desired option to copy it to the Mode screen. Then press $\boxed{\text{ENTER}}$ to save it there. Instead of pressing the number of the desired setting, you can highlight it and then press $\boxed{\text{ENTER}}$ $\boxed{\text{ENTER}}$ to copy it to the Mode screen and save it there. Note that the cursor skips dimmed settings as you move through the options.

It will be helpful to read Chapter 1: Getting Started and Chapter 2: Operating the TI-89 in the TI-89 Guidebook before proceeding.

USING A MENU

A menu is a list of options that appear when a key is pressed. For example, press $\boxed{\text{F1}}$ to display the Tools menu. We can select an item from a menu by using ∇ to highlight it and then pressing $\boxed{\text{ENTER}}$ or by simply pressing the number of the item. If we press 8 when the Tools menu is displayed, for instance, we select "Clear Home" and any previously entered computations will be cleared from the history area of the home screen. If an item is identified by a letter rather than a number, press the purple $\boxed{\text{alpha}}$ key followed by the letter of the item to select it. The letters are printed in purple above the keys on the keypad. The down-arrow beside item 8 in the menu below indicates that there are additional items in the menu. Use ∇ to scroll down to them.

SETTING THE VIEWING WINDOW

Section 1.1, page 7 (Page numbers refer to pages in the textbook.) The viewing window is the portion of the coordinate plane that appears on the grapher's screen. It is defined by the minimum and maximum values of x and y: xmin, xmax, ymin, and ymax. The notation [xmin, xmax, ymin, ymax] is used in the text to represent these window settings or dimensions. For example, $[-12, 12, -8, 8]$ denotes a window that displays the portion of the x-axis from -12 to 12 and the portion of the y-axis from -8 to 8. In addition, the distance between tick marks on the axes is defined by the settings xscl and yscl. In this manual xscl and yscl will be assumed to be 1 unless noted otherwise. The setting xres sets the pixel resolution. We usually select xres = 2. The window corresponding to the settings $[-20, 30, -12, 20]$, xscl = 5, yscl = 2, xres = 2, is shown below.

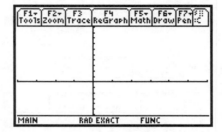

Press ◇ WINDOW to display the current window settings on your grapher. (WINDOW is the green ◇ operation associated with the F2 key on the top row of the keypad.) The standard settings are shown below.

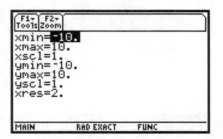

To change a setting, position the cursor beside the setting you wish to change and enter the new value. For example, to change from the standard settings to $[-20, 30, -12, 20]$, xscl = 5, yscl = 2, on the WINDOW screen, start with the setting beside "xmin =" highlighted and press (−) 2 0 ENTER 3 0 ENTER 5 ENTER (−) 1 2 ENTER 2 0 ENTER 2 ENTER . You must use the (−) key on the bottom row of the keypad rather than the − key in the right-hand column to enter a negative number. (−) represents "the opposite of" or "the additive inverse of" whereas − is the key for the subtraction operation. The ▽ key may be used instead of ENTER after typing each window setting. To see the window shown above, press ◇ GRAPH . (GRAPH is the green ◇ operation associated with the F3 key on the top row of the keypad.)

QUICK TIP: To return quickly to the standard window setting $[-10, 10, -10, 10]$, xscl = 1, yscl = 1, when either the Window screen or the Graph screen is displayed, press F2 to access the ZOOM menu and then press 6 to select item 6,

ZoomStd (Zoom Standard).

GRAPHS

After entering an equation and setting a viewing window, you can view the graph of an equation.

Section 1.1, page 8 Graph $y = x^3 - 5x + 1$ using a graphing calculator.

Equations are entered on the equation-editor screen. Press $\boxed{\diamond}$ $\boxed{Y=}$ to access this screen. ($Y =$ is the green $\diamond$ operation associated with the $\boxed{F1}$ key.) If any plots are turned on they should be turned off, or deselected, now. A check mark beside the name of a plot indicates that it is currently selected. To deselect it, move the cursor to the plot. Then press $\boxed{F4}$. There should now be no check mark beside the name of the plot, indicating that it has been deselected. If there is currently an expression displayed for y_1, clear it by positioning the cursor beside "y1 =" and then press $\boxed{\text{CLEAR}}$. Do the same for expressions that appear on all other "y =" lines by using $\boxed{\triangledown}$ to move to a line and then pressing $\boxed{\text{CLEAR}}$. Then use $\boxed{\triangle}$ or $\boxed{\triangledown}$ to move the cursor beside "y1 =." Now enter $y_1 = x^3 - 5x + 1$ on the entry line of the equation-editor screen and paste it beside "y1 =" by pressing $\boxed{X}$ $\boxed{\wedge}$ $\boxed{3}$ $\boxed{-}$ $\boxed{5}$ $\boxed{X}$ $\boxed{+}$ $\boxed{1}$ $\boxed{\text{ENTER}}$.

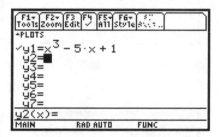

The standard $[-10, 10, -10, 10]$ window is a good choice for this graph. Either enter these dimensions in the WINDOW screen and then press $\boxed{\diamond}$ $\boxed{\text{GRAPH}}$ to see the graph or simply press $\boxed{F2}$ 6 to select the standard window and see the graph.

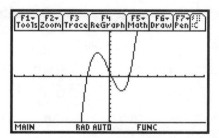

To edit an entry on the equation-editor screen, use $\boxed{\triangle}$ or $\boxed{\triangledown}$ to highlight it and then press $\boxed{\text{ENTER}}$. This copies the entry to the entry line where it can be edited. If, for instance, in the expression above you pressed 6 instead of 5, first press $\boxed{\triangleleft}$ to move the cursor toward the 6. Now, to type a 5 over the 6, the grapher must be set in overtype mode. If you see a thick, blinking cursor that highlights the entire character, your grapher is in this mode. If you see a thin, vertical cursor positioned between two characters, your grapher is in insert mode. Select overtype mode by pressing $\boxed{\text{2nd}}$ $\boxed{\text{INS}}$.

(INS is the second operation associated with the $\boxed{\leftarrow}$ key.) Now the cursor becomes a dark, blinking rectangle rather than a vertical line. Use $\boxed{\triangleright}$ to position the cursor over the 6 and then press 5 to write a 5 over the 6. To leave overtype mode press $\boxed{\text{2nd}}$ $\boxed{\text{INS}}$ again. The grapher is now in the insert mode, indicated by a vertical cursor, and will remain in that mode until overtype mode is once again selected.

If you forgot to type the plus sign, move the insert cursor to the left of the 1 and press $\boxed{+}$ to insert the plus sign before the 1. You can continue to insert symbols immediately after the first insertion. If you typed 52 instead of 5, move the cursor to the right of 2 and press $\boxed{\leftarrow}$. This will delete the 2. Instead of using overtype mode to overtype a character as described above, we can use $\boxed{\leftarrow}$ to delete the character and then, in insert mode, insert a new character.

An equation must be solved for y before it can be graphed on the TI-89.

Section 1.1, page 8 To graph $3x + 5y = 10$, first solve for y, obtaining $y = \dfrac{-3x + 10}{5}$. Then press $\boxed{\diamond}$ $\boxed{\text{Y} =}$ and clear any expressions that currently appear. Position the cursor beside "y1 =." Now press $\boxed{(}$ $\boxed{(-)}$ 3 $\boxed{\text{X}}$ $\boxed{+}$ 1 0 $\boxed{)}$ $\boxed{\div}$ 5 to enter the right-hand side of the equation. Note that without the parentheses the expression $-3x + \dfrac{10}{5}$ would have been entered.

Select a viewing window and then press $\boxed{\diamond}$ $\boxed{\text{GRAPH}}$ to display the graph. You may change the viewing window as desired to reveal more or less of the graph. The standard window is shown here.

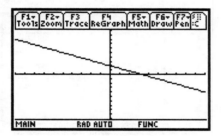

To graph $x = y^2$, first solve the equation for y : $y = \pm\sqrt{x}$. To obtain the entire graph of $x = y^2$, you must graph $y_1 = \sqrt{x}$ and $y_2 = -\sqrt{x}$ on the same screen. Press $\boxed{\diamond}$ $\boxed{\text{Y} =}$ and clear any expressions that currently appear. With the cursor beside "y1 =" press $\boxed{\text{2nd}}$ $\boxed{\sqrt{}}$ $\boxed{\text{X}}$ $\boxed{)}$ $\boxed{\text{ENTER}}$. ($\boxed{\sqrt{}}$ is the second operation associated with the $\boxed{\times}$ multiplication key.) On the TI-89, a left parenthesis appears along with the radical symbol, so a separate keystroke is not necessary to introduce it.

Now use $\boxed{\triangledown}$ to move the cursor beside "y2 =." We will show two ways to enter $y_2 = -\sqrt{x}$. One is to enter the expression $-\sqrt{x}$ directly by pressing $\boxed{(-)}$ $\boxed{\text{2nd}}$ $\boxed{\sqrt{}}$ $\boxed{\text{X}}$ $\boxed{)}$ $\boxed{\text{ENTER}}$.

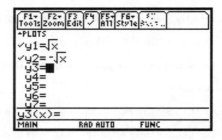

The other method of entering y_2 is based on the observation that $-\sqrt{x}$ is the opposite of the expression for y_1. That is, $y_2 = -y_1$, so we can enter $y_2 = -y_1(x)$. To do this, press $\boxed{(-)}$ $\boxed{Y}$ $\boxed{1}$ $\boxed{(}$ $\boxed{X}$ $\boxed{)}$ $\boxed{\text{ENTER}}$.

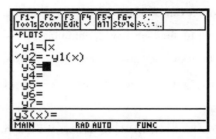

Select a viewing window and press $\boxed{\text{GRAPH}}$ to display the graph. The window shown here is $[-2, 10, -5, 5]$.

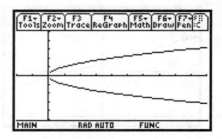

The top half is the graph of y_1, the bottom half is the graph of y_2, and together they yield the graph of $x = y^2$.

THE TABLE FEATURE

For an equation entered in the equation-editor screen, a table of x-and y-values can be displayed.

Section 1.2, page 18 Create a table of ordered pairs for the function $f(x) = x^3 - 5x + 1$.

Enter the function as $y_1 = x^3 - 5x + 1$ as described on page 120 of this manual. Be sure that plots are turned off and that any previous entries are cleared. (See page 120 of this manual for the procedure.) Once the equation in entered, press $\boxed{\diamond}$ $\boxed{\text{TblSet}}$ or $\boxed{\diamond}$ $\boxed{\text{TABLE}}$ $\boxed{\text{F2}}$ to access the TABLE SETUP window. (TblSet is the green $\diamond$ operation associated with the $\boxed{\text{F4}}$ key.) If "Independent" is set to "Auto" on the Table Setup screen, the grapher will supply values for x, beginning with the value specified as tblStart and continuing by adding the value of Δtbl to the preceding value for x. If the table was previously set to Ask, the blinking cursor will be positioned over ASK. Change this setting to AUTO by pressing $\boxed{\triangleright}$ 1. Now use the $\boxed{\triangle}$ key to move the cursor to tblStart. Enter a minimum x-value of 0.3, an increment of 1, and a Graph $< - >$ Table setting of OFF by first positioning the cursor beside tblStart and then pressing $\boxed{}$ $\boxed{.}$ 3 $\boxed{\bigtriangledown}$ 1 $\boxed{\bigtriangledown}$ $\boxed{\triangleright}$ 1 $\boxed{\text{ENTER}}$. Press $\boxed{\diamond}$ $\boxed{\text{TABLE}}$ to see the table. (TABLE is the green $\diamond$ operation associated with the $\boxed{\text{F5}}$ key.)

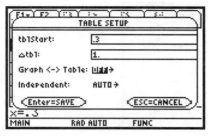

Use the $\boxed{\bigtriangledown}$ and $\boxed{\triangle}$ keys to scroll through the table. For example, by using $\boxed{\bigtriangledown}$ to scroll down we can see that $y_1 = 758.9$ when $x = 9.3$. Using $\boxed{\triangle}$ to scroll up, observe that $y_1 = -31.2$ when $x = -3.7$.

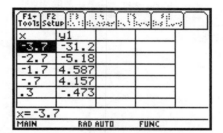

GRAPHS AND FUNCTION VALUES

Section 1.2, page 21 There are several ways to evaluate a function using a grapher. Three of them are described here. Given the function $f(x) = 2x^2 + x$, we will find $f(-2)$. First press $\boxed{\diamond}$ $\boxed{\text{Y} =}$ and enter the function as $y_1 = 2x^2 + x$. Now we will find $f(-2)$ using the TABLE feature. Press $\boxed{\diamond}$ $\boxed{\text{TblSet}}$ or $\boxed{\diamond}$ $\boxed{\text{TABLE}}$ $\boxed{\text{F2}}$ to access the TableSetup screen. Move the cursor to the "Independent" line. Then press $\boxed{\triangleright}$ 2 $\boxed{\text{ENTER}}$ to select Ask mode. In Ask mode the grapher disregards the other settings on the Table Setup screen.

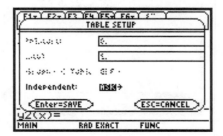

Now press $\boxed{\diamond}$ $\boxed{\text{TABLE}}$ to view the table. (TABLE is the second operation associated with the $\boxed{\text{F5}}$ key.) If you select Ask before a table is displayed for the first time on your grapher, a blank table is displayed. If a table has previously been displayed, the table you now see will continue to show the values in the previous table.

Values for x can be entered in the x-column of the table and the corresponding values for y_1 will be displayed in the

$y1$-column. To enter -2, for instance, press $\boxed{(-)}$ 2 $\boxed{\text{ENTER}}$. Any additional x-values that are displayed are from a table that was previously displayed on the Auto setting.

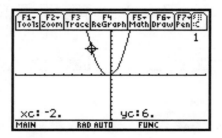

We see that $y_1 = 6$ when $x = -2$, so $f(-2) = 6$.

We can also use the Value feature from the Math menu on the Graph screen to find $f(-2)$. To do this, graph $y_1 = 2x^2 + x$ in a window that includes the x-value -2. We will use the standard window. Then press $\boxed{\text{F5}}$ 1 to select the Value feature. Now supply the desired x-value by pressing $\boxed{(-)}$ 2. Press $\boxed{\text{ENTER}}$ to see xc$:= -2$ and yc:6 at the bottom of the screen. Thus $f(-2) = 6$.

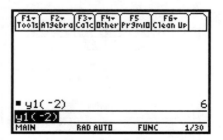

A third method for finding $f(-2)$ uses function notation directly. With $y_1 = 2x^2 + x$ entered on the Y = screen, go to the home screen by pressing $\boxed{\text{HOME}}$ or $\boxed{\text{2nd}}$ $\boxed{\text{QUIT}}$. (QUIT is the second operation associated with the $\boxed{\text{ESC}}$ key.) Now enter y1(-2) by pressing $\boxed{\text{Y}}$ 1 $\boxed{(}$ $\boxed{(-)}$ 2 $\boxed{)}$ $\boxed{\text{ENTER}}$. Again we see that $y_1(-2) = 6$, or $f(-2) = 6$.

GRAPHING FUNCTIONS DEFINED PIECEWISE

Section 1.2, Example 9, page 22   Graph: $f(x) = \begin{cases} 4 \text{ for } x \leq 0, \\ 4 - x^2 \text{ for } 0 < x \leq 2, \\ 2x - 6 \text{ for } x > 2. \end{cases}$

We will create a multi-statement, user-defined function to graph this function. It is presented here first in block form.

Func

   If x<=0 Then

    Return 4

  ElseIf x>0 and x<2 Then

    Return 4-x^2

  Else

    Return 2x-6

  EndIf

  EndFunc

When this function is entered on the equation-editor screen, the entire function must be entered on a single line with a colon used to separate statements. That is, we enter Func:If x<=0 Then:Return 4: . . . :EndIf:EndFunc.

Letters must be capitalized and spaces used exactly as shown above. To type a capital letter press $\boxed{\uparrow}$ followed by the letter key. To type a single lowercase letter, press $\boxed{\text{alpha}}$ and then the letter key. To turn on the lowercase alpha-lock press $\boxed{\text{2nd}}$ $\boxed{\text{a-lock}}$. (a-lock is the second operation associated with the $\boxed{\text{alpha}}$ key.) Press $\boxed{\text{alpha}}$ to turn off the alpha-lock. The : is the second operation associated with the 4 numeric key. Enter a space by pressing $\boxed{\text{alpha}}$ and then the $\boxed{(-)}$ key.

After the function has been entered, the equation-editor screen will display "y1 = Func."

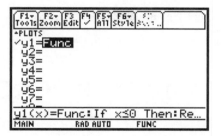

Now select the Dot graph style. If this is not done, a vertical line that is not part of the graph will appear. Dot style can be selected from the Style menu by highlighting Func on the equation-editor screen and pressing $\boxed{\text{2nd}}$ $\boxed{\text{F6}}$ 2 or $\boxed{\text{2nd}}$ $\boxed{\text{F6}}$ $\boxed{\triangledown}$ $\boxed{\text{ENTER}}$. Choose and enter window dimensions and then press $\boxed{\diamond}$ $\boxed{\text{GRAPH}}$ to see the graph of the function. It is shown here in the window $[-5, 5, -3, 6]$.

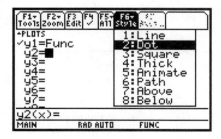 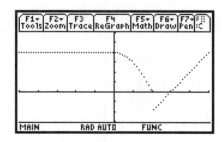

THE TRACE FEATURE

The TRACE feature can be used to display the coordinates of points on a graph.

Section 1.2, page 25 Enter the function $f(x) = x^3 - 5x + 1$ (see page 120 of this manual) and graph it in the window $[-5, 5, -10, 10]$. Now, from the Graph screen, press $\boxed{\text{F3}}$ to activate the TRACE feature. The trace cursor appears on the graph and the coordinates of the point at which it is positioned are displayed at the bottom of the screen. Use the $\boxed{\triangleleft}$ and $\boxed{\triangleright}$ keys to move the cursor along the graph to see the coordinates of other points.

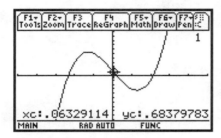
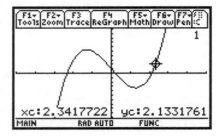

SQUARING THE VIEWING WINDOW

Section 1.4, page 41 In the standard window, the distance between tick marks on the y-axis is about $1/2$ the distance between tick marks on the x-axis. It is often desirable to choose window dimensions for which these distances are the same, creating a "square" window. Any window in which the ratio of the length of the y-axis to the length of the x-axis is $1/2$ will produce this effect. This can be accomplished by selecting dimensions for which $\text{ymax} - \text{ymin} = \dfrac{1}{2}(\text{xmax} - \text{xmin})$.

The standard window is shown on the left below and the square window $[-6, 6, -3, 3]$ is shown on the right. Observe that the distance between tick marks appears to be the same on both axes in the square window.

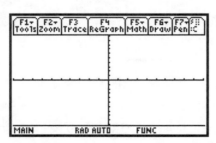
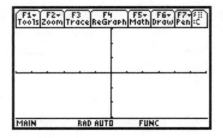

The window can also be squared by using the grapher's ZoomSqr feature. From the equation-editor, Window, or Graph screen, press $\boxed{\text{F2}}$ 5 to select the ZoomSqr window. Starting with the standard window and pressing $\boxed{\text{F2}}$ 5 produces the dimensions and the window shown below.

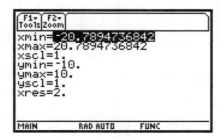
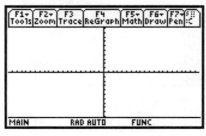

THE INTERSECTION FEATURE

We can use the Intersection feature from the CALC menu to solve equations.

Section 1.5, page 61 Solve the equation $x^3 = 3x + 1$ using the Intersection feature.

On the equation-editor screen, clear any existing entries and then enter $y_1 = x^3$ and $y_2 = 3x + 1$. Now graph these equations in an appropriate window. One good choice is $[-3, 3, -10, 10]$. The solutions of the equation $x^3 = 3x + 1$ are the first coordinates of the points of intersection of these graphs. We will use the Intersection feature to find the leftmost point of intersection first. Press $\boxed{\text{F5}}$ 5 to select Intersection from the Math menu on the Graph screen. The query "1st curve?" appears at the bottom of the screen. The blinking cursor is positioned on the graph of y_1. This is indicated by the 1 in the upper right-hand corner of the screen. Press $\boxed{\text{ENTER}}$ to indicate that this is the first curve involved in the intersection. Next the query "2nd curve?" appears at the bottom of the screen. The blinking cursor is now positioned on the graph of y_2 and the notation 2 should appear in the top right-hand corner of the screen. Press $\boxed{\text{ENTER}}$ to indicate that this is the second curve. We identify the curves for the grapher since we could have more than two graphs on the screen at once. After we identify the second curve, the query "Lower bound?" appears at the bottom of the screen. Use the right and left arrow keys to move the blinking cursor to a point to the left of the point of intersection of the lines or type an x-value less than the x-coordinate of the point of intersection. Then press $\boxed{\text{ENTER}}$. Next the query "Upper bound?" appears. We give a lower and an upper bound since some pairs of curves have more than one point of intersection. Move the cursor to a point to the right of the point of intersection or type an x-value greater than the x-value of the point of intersection and press $\boxed{\text{ENTER}}$. Now the coordinates of the point of intersection appear at the bottom of the screen.

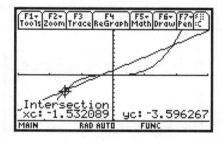

We see that, at the leftmost point of intersection, $x \approx -1.53$, so one solution of the equation is approximately -1.53. Repeat this process two times to find the coordinates of the other two points of intersection. We find that the other two solutions of the equation are approximately -0.35 and 1.88.

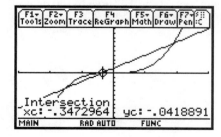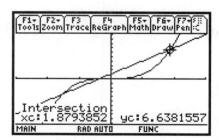

THE ZERO FEATURE

When an equation is expressed in the form $f(x) = 0$, it can be solved using the Zero feature from the Math menu on the Graph screen.

Section 1.5, page 61 Solve the equation $x^3 = 3x + 1$ using the Zero feature.

First subtract $3x$ and 1 on both sides of the equation to obtain an equivalent equation with 0 on one side. We have $x^3 - 3x - 1 = 0$. The solutions of the equation $x^3 = 3x + 1$ are the values of x for which the function $f(x) = x^3 - 3x - 1$ is equal to 0. We can use the Zero feature to find these values, or zeros.

On the equation-editor screen, clear any existing entries and then enter $y_1 = x^3 - 3x - 1$. Now graph the function in a viewing window that shows the x-intercepts clearly. One good choice is $[-3, 3, -5, 8]$. We see that the function has three zeros. They appear to be about -1.5, -0.5, and 2.

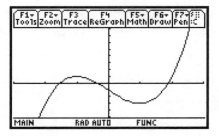

We will find the zero near -1.5 first. Press F5 2 to select the Zero feature from the Math menu. We are prompted to select a lower bound. This means that we must choose an x-value that is to the left of -1.5 on the x-axis. This can be done by using the left- and right-arrow keys to move to a point on the curve to the left of -1.5 or by keying in a value less than -1.5.

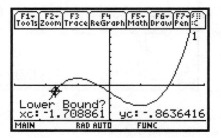

Once this is done press ENTER. Now we are prompted to select an upper bound that is to the right of -1.5 on the x-axis. Again, this can be done by using the arrow keys to move to a point on the curve to the right of -1.5 or by keying in a value greater that -1.5.

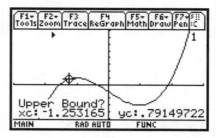

Press ENTER again. We see that $y = 0$ when $x \approx -1.53$, so -1.53 is a zero of the function f.

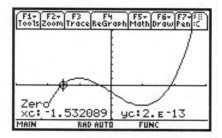

Select Zero from the Math menu a second time to find the zero near -0.5 and a third time to find the zero near 2. We see that the other two zeros are approximately -0.35 and 1.88.

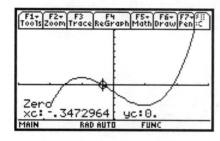 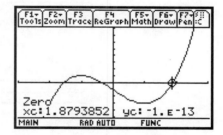

ABSOLUTE-VALUE FUNCTIONS

We can use the absolute-value option to perform computations involving absolute value and to graph absolute-value functions.

Section 1.5, page 65 Graph $f(x) = |x|$.

The absolute-value option is accessed from the catalog. To enter $y = |x|$, first press ◇ F1 to go to the equation-editor screen and then clear any existing entries. Now position the cursor beside y1 = and enter $|x|$.

To do this, first press CATALOG, position the cursor beside "abs," and the press ENTER to copy "abs" to the equation-editor screen. To go to "abs(" quickly, press A, the purple alpha operation associated with the = key. Now press X) ENTER. Choose an appropriate viewing window and graph the function. To use the standard window, press F2 6.

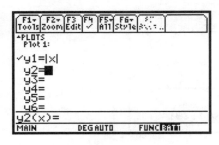 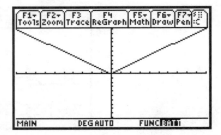

GRAPHING RADICAL FUNCTIONS

There are various way to enter radical expressions on the TI-89.

Section 1.5, page 67 We discussed entering an expression containing a square root on page 121 of this manual. Remember that a left parenthesis appears along with the radical symbol. We must close the parentheses after the radicand is entered.

Higher order radical expressions can be entered using rational exponents. To enter $y_1 = \sqrt[3]{x-2}$ as $y_1 = (x-2)^{1/3}$, for example, position the cursor beside y1 = and press $\boxed{(}$ $\boxed{X}$ $\boxed{-}$ $\boxed{2}$ $\boxed{)}$ $\boxed{\wedge}$ $\boxed{(}$ $\boxed{1}$ $\boxed{\div}$ $\boxed{3}$ $\boxed{)}$ $\boxed{\text{ENTER}}$. To enter $y_2 = \sqrt[5]{6-x}$ as $y_2 = (6-x)^{1/5}$ press $\boxed{(}$ $\boxed{6}$ $\boxed{-}$ $\boxed{X}$ $\boxed{)}$ $\boxed{\wedge}$ $\boxed{(}$ $\boxed{1}$ $\boxed{\div}$ $\boxed{5}$ $\boxed{)}$ $\boxed{\text{ENTER}}$.

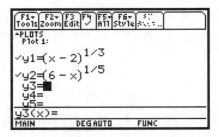

LINEAR REGRESSION

We can use the Linear Regression feature to fit a linear equation to a set of data.

Section 1.6, page 79 The following table lists data showing the price P of a one-day adult admission to Disney World for years since 1993.

Years, x (since 1993)	Price P of a One-Day Adult Admission to Disney World
0. 1993	$34.00
1. 1994	$36.00
2. 1995	$37.00
3. 1996	$40.81
4. 1997	$42.14
5. 1998	$44.52

(a) Fit a regression line to the data using the REGRESSION feature on a grapher.

(b) Graph the regression line with the scatterplot.

(c) Use the model to predict the price of a one-day adult admission in 2000 ($x = 7$).

(a) We will enter the data points in the Data/Matrix editor. Press $\boxed{\text{APPS}}$ 6 3 to display a new data variable screen in the Data/Matrix editor. We must now enter a data variable name in the Variable box on this screen. The name can contain from 1 to 8 characters and cannot start with a numeral. Some names are preassigned to other uses on the TI-89. If you try to use one of these, you will get an error message. Press $\boxed{\bigtriangledown}$ $\boxed{\bigtriangledown}$ to move the cursor to the Variable box. We will name our data variable "price." To enter this name, first lock the alphabetic keys on by pressing $\boxed{\text{2nd}}$ $\boxed{\text{a-lock}}$. Then press $\boxed{\text{P}}$ $\boxed{\text{R}}$ $\boxed{\text{I}}$ $\boxed{\text{C}}$ $\boxed{\text{E}}$. Note that P, R, I, C, and E are the purple alphabetic operations associated with the $\boxed{\text{STO} \triangleright}$, 2, 9, $\boxed{)}$, and $\boxed{\div}$ keys, respectively.

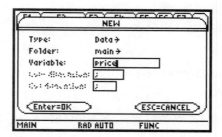

After typing the name of the data variable, unlock the alphabetic keys by pressing the purple $\boxed{\text{alpha}}$ key. Now press $\boxed{\text{ENTER}}$ $\boxed{\text{ENTER}}$ to go to the data-entry screen. Assuming the data variable name "price" has not previously been used in your calculator, this screen will contain empty data lists with row 1, column 1 highlighted. If entries have previously been made in a data variable named "price," they can be cleared by pressing $\boxed{\text{F1}}$ 8 $\boxed{\text{ENTER}}$.

We will enter the first coordinates (x-coordinates) of the points in column c1 and the second coordinates (y-coordinates) in c2. To enter the first x-coordinate, 0, press 0 $\boxed{\text{ENTER}}$. Continue typing the x-values 1, 2, 3, 4, and 5, each followed by $\boxed{\text{ENTER}}$. The entries can be followed by $\boxed{\bigtriangledown}$ rather than $\boxed{\text{ENTER}}$ if desired. Press $\boxed{\triangleright}$ $\boxed{\bigtriangleup}$ $\boxed{\bigtriangleup}$ $\boxed{\bigtriangleup}$ $\boxed{\bigtriangleup}$ $\boxed{\bigtriangleup}$ $\boxed{\bigtriangleup}$ to move to the top of column c2. Type the y-values 34, 36, 37, 40.81, 42.14, and 44.52 in succession, each followed by $\boxed{\text{ENTER}}$ or $\boxed{\bigtriangledown}$. Note that the coordinates of each point must be in the same position in both lists.

Now press $\boxed{\text{F5}}$ to display the Calculate menu. Press $\boxed{\triangleright}$ 5 to select LinReg (linear regression). Then press $\boxed{\bigtriangledown}$ $\boxed{\text{alpha}}$ $\boxed{\text{C}}$ 1 $\boxed{\bigtriangledown}$ $\boxed{\text{alpha}}$ $\boxed{\text{C}}$ 2 to indicate that the data in c1 and c2 will be used for x and y, respectively. Press $\boxed{\bigtriangledown}$ $\boxed{\triangleright}$ $\boxed{\bigtriangledown}$ $\boxed{\text{ENTER}}$ to indicate that the regression equation should be copied to the equation-editor screen as y_1. Finally press $\boxed{\text{ENTER}}$ again to see the STAT VARS screen which displays the coefficients a and b of the regression equation $y = ax + b$.

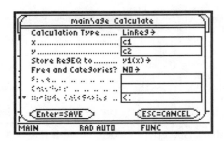 

Note that values for "corr" (the correlation coefficient) and r^2 (the coefficient of determination) will also be displayed. These numbers indicate how well the regression line fits the data. While it is possible to suppress these numbers on some graphers, this cannot be done on the TI-89.

(b) To plot the data points we access the Plot Setup screen from the Data/Matrix editor by pressing $\boxed{\text{APPS}}$ 6 1 $\boxed{\text{F2}}$. We will use Plot 1, which is highlighted. If any plot settings are currently entered beside "Plot 1," clear them by pressing $\boxed{\text{F3}}$. Clear settings shown beside any other plots as well by using $\boxed{\bigtriangledown}$ to highlight each plot in turn and then pressing $\boxed{\text{F3}}$.

Now we define Plot 1. Use $\boxed{\triangle}$ to highlight Plot 1 if necessary. Then press $\boxed{\text{F1}}$ to display the Plot Definition screen. The item on the first line, Plot Type, is highlighted. We will choose a scatterplot, by pressing $\boxed{\triangleright}$ 1. Now press $\boxed{\bigtriangledown}$ to go to the next line, Mark. Here we select the type of mark or symbol that will be used to plot the points. We select a box by pressing $\boxed{\triangleright}$ 1. Now we must tell the grapher which columns of the data variable to use for the x- and y-coordinates of the points to be plotted. Press $\boxed{\bigtriangledown}$ to move the cursor to the "x" line and enter c1 as the source of the x-coordinates by pressing $\boxed{\text{alpha}}$ $\boxed{\text{C}}$ 1. (C is the purple alphabetic operation associated with the $\boxed{)}$ key.) Press $\boxed{\bigtriangledown}$ $\boxed{\text{alpha}}$ $\boxed{\text{C}}$ 2 to go the the "y" line and enter c2 as the source of the y-coordinates.

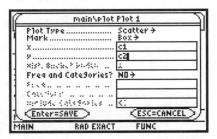

Save the plot definition and return to the Plot Setup screen by pressing $\boxed{\text{ENTER}}$ $\boxed{\text{ENTER}}$. Beside "Plot 1:" you will now see a shorthand notation for the definition of the plot.

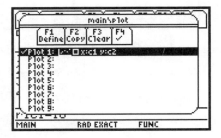

Now press ◇ WINDOW to go to the Window screen and select a viewing window. The years range from 0 through 5 and the prices range from $34.00 through $44.52, so one good choice is $[-1, 6, 30, 50]$.

Recall that when we found the regression equation in part (a) we also copied it to the equation-editor screen as y_1. To see the plotted points and the regression line, press ◇ GRAPH .

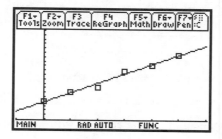

QUICK TIP: Instead of entering the window dimensions directly, we can press F2 9 after entering the coordinates of the points in lists and defining Plot 1. This activates the ZoomData operation which automatically defines a viewing window that displays all of the points and also displays the graph.

(c) To predict the price of a one-day adult admission in 2000, evaluate the regression equation for $x = 7$. (2000 is 7 years after 1993.) Use any of the methods for evaluating a function presented earlier in this chapter. (See pages 123 and 124.) We will use function notation on the home screen.

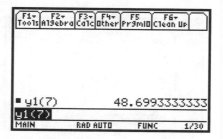

When $x = 7, y \approx 48.70$, so we predict that the price of a one-day adult admission to Disney World will be about $48.70 in 2000.

POLYNOMIAL REGRESSION

The TI-89 has the capability to use regression to fit quadratic, cubic, and quartic functions to data.

Section 1.6, page 81 The following chart relates the number of live births to women of a particular age.

Age, x	Average Number of Live Births per 1000 Women
16	34
18.5	86.5
22	111.1
27	113.9
32	84.5
37	35.4
42	6.8

(a) Fit a quadratic function to the data using the REGRESSION feature on a grapher.

(b) Make a scatterplot of the data. Then graph the quadratic function with the scatterplot.

(c) Fit a cubic function to the data using the REGRESSION feature on a grapher.

(d) Make a scatterplot of the data. Then graph the cubic function with the scatterplot.

(e) Decide which function seems to fit the data better.

(f) Use the function from part (e) to estimate the average number of live births by women of ages 20 and 30.

(a) First enter the data in the Data/Matrix editor as described on page 131 of this manual. Then select QuadReg as the CalculationType from the Calculate menu by pressing F5 ▷ 9. Specify the sources of x and y. We will also copy the equation to the equation-editor screen so that we can graph it. (See page 131 of this manual for the procedure.) Then press ENTER ENTER. The grapher returns the coefficients of a quadratic function $y = ax^2 + bx + c$. Rounding the coefficients to two decimal places, we obtain the function $f(x) = -0.49x^2 + 25.95x - 238.49$.

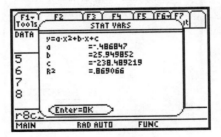

(b) Graph the function along with the scatterplot as described on pages 132 and 133 of this manual.

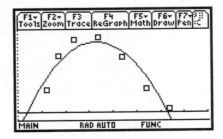

(c) Once the data are entered, fit a cubic function to it by selecting cubic regression, denoted CubicReg, from the Calculate menu. This is item 3 under CalculationType. Note that we will copy the equation to the equation-editor screen as well so that we can graph it. (See page 131 for the procedure.) The grapher returns the coefficients for a cubic function of the form

$h(x) = ax^3 + bx^2 + cx + d$. Rounding the coefficients to two decimal places, we have $f(x) = 0.03x^3 - 3.22x^2 + 101.18x - 886.93$.

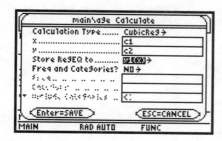 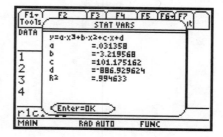

(d) Graph the function along with the scatterplot as described on pages 132 and 133 of this manual.

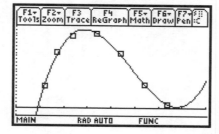

(e) The graph of the cubic function appears to fit the data better than the graph of the quadratic function.

(f) We can use any of the methods described earlier in this chapter to evaluate the function. We will use function notation on the home screen.

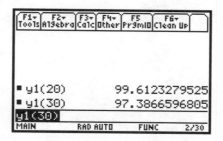

We estimate that there will be about 100 live births per 1000 20-year-old women and about 97 live births per 1000 30-year-old women.

A quartic function can be fit to data by selecting quartic regression, denoted QuartReg, from the Calculate menu. The grapher returns the coefficients of a function $f(x) = ax^4 + bx^3 + cx^2 + dx + e$.

Chapter 2
Differentiation

ZOOMING IN

The Zoom In operation from the ZOOM menu can be used to enlarge a portion of a graph.

Section 2.2, page 112 Verify graphically that $\lim\limits_{x \to 0} \dfrac{\sqrt{x+1}-1}{x} = 0.5$.

First graph $y = \dfrac{\sqrt{x+1}-1}{x}$ in a window that shows the portion of the graph near $x = 0$. One good choice is $[-2, 5, -1, 2]$. Then press $\boxed{\text{F3}}$ to select Trace and use the $\boxed{\triangleleft}$ and $\boxed{\triangleright}$ keys to move the trace cursor to a point on the curve near $x = 0$.

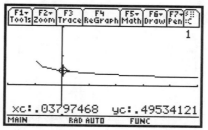

Now select the Zoom In operation from the ZOOM menu by pressing $\boxed{\text{F2}}$ $\boxed{2}$ $\boxed{\text{ENTER}}$. This enlarges the portion of the graph near $x = 0$. We can now press $\boxed{\text{F3}}$ again and use the $\boxed{\triangleleft}$ and $\boxed{\triangleright}$ keys to trace the curve near $x = 0$.

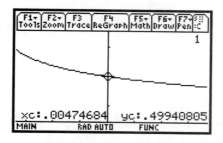 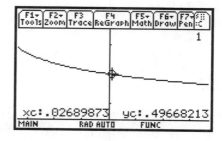

The Zoom In operation can be used as many times as desired in order to verify the result.

THE nDeriv OPERATION

The TI-89 can be used to find the slope of a line tangent to a curve at a specific point. That is, it can find the derivative of a function $f(x)$ for a specific value of x.

Section 2.4, page 134 For the function $f(x) = x(100 - x)$, find $f'(70)$.

We will use the nDeriv (numerical derivative) operation. This is item A on the Calc menu.

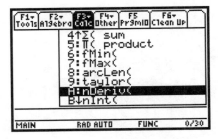

Select this operation from the home screen by pressing $\boxed{\text{F3}}$ $\boxed{\text{alpha}}$ $\boxed{\text{A}}$ or by pressing $\boxed{\text{F3}}$, using the $\boxed{\triangledown}$ key to highlight item A, and then pressing $\boxed{\text{ENTER}}$. These keystrokes copy "nDeriv(" to the entry line of the home screen. Now enter the function, the variable, and the value at which the derivative is to be evaluated. Note that we must include a multiplication symbol when entering the function: $x \times (100 - x)$. Press $\boxed{\text{X}}$ $\boxed{\times}$ $\boxed{(}$ 1 0 0 $\boxed{-}$ $\boxed{\text{X}}$ $\boxed{)}$ $\boxed{,}$ $\boxed{\text{X}}$ $\boxed{)}$ $\boxed{|}$ $\boxed{\text{X}}$ $\boxed{=}$ 7 0 $\boxed{\text{ENTER}}$. Note that the grapher supplies a left parenthesis after "nDeriv" and we close the parentheses with a right parenthesis after entering the variable, x. We see that $f'(70) = -40$.

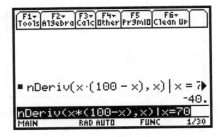

THE TANGENT FEATURE

The TI-89 does not have a feature equivalent to the Tangent feature described in **Section 2.4, page 134**. See page 139 of this manual for a procedure that draws a tangent line directly from the GRAPH screen.

THE dy/dx OPERATION

We can find the derivative of a function at a specific point directly from the GRAPH screen.

Section 2.5, page 139 For the function $f(x) = x\sqrt{4 - x^2}$, find dy/dx at a specific point.

First graph $y = x\sqrt{4 - x^2}$. We will use the window $[-3, 3, -4, 4]$. Then select the dy/dx operation from the Derivatives submenu of the Math menu on the Graph screen by pressing $\boxed{\text{F5}}$ 6 $\boxed{\text{ENTER}}$. We see the graph with a cursor positioned on it near the middle of the window.

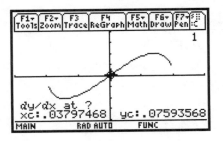

To find the value of dy/dx at a specific point either move the cursor to the desired point or key in its x-coordinate. For example, if we move the cursor to the point (1.2531646, 1.9533208) and press $\boxed{\text{ENTER}}$ we find that $dy/dx = 0.55119739$ at this point.

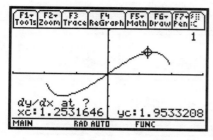 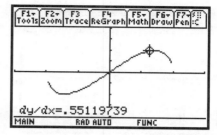

To find dy/dx for $x = 1$, select dy/dx from the Derivatives submenu of the Math menu and then press 1 $\boxed{\text{ENTER}}$. We see that $dy/dx = 1.1547005$ at this point.

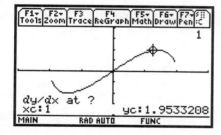 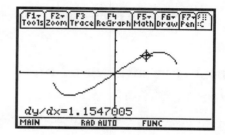

MORE ON TANGENT LINES

We can draw a line tangent to a curve at a specific point directly from the GRAPH screen.

Section 2.5, page 139 Draw the line tangent to the graph of $f(x) = x\sqrt{4 - x^2}$ at a specific point.

First graph $y = x\sqrt{4 - x^2}$ in an appropriate window such as $[-3, 3, -4, 4]$. Then press $\boxed{\text{TRACE}}$ and use the $\boxed{\triangleleft}$ and $\boxed{\triangleright}$ keys to move the cursor to the desired point. Now select the Tangent feature from the Math menu on the Graph screen by pressing $\boxed{\text{F5}}$ $\boxed{\text{alpha}}$ $\boxed{\text{A}}$ or by pressing $\boxed{\text{F5}}$, using $\boxed{\triangledown}$ to highlight item A, and pressing $\boxed{\text{ENTER}}$. Press $\boxed{\text{ENTER}}$ to see the tangent line at this point and its equation.

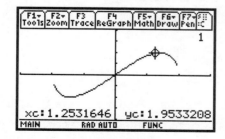 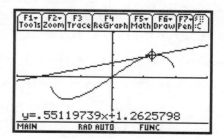

To clear the graph of the tangent line from the Graph screen press $\boxed{\text{F4}}$, ReGraph. We can also use the ClrDraw (clear drawing) operation from the Draw menu on the Graph screen. To do this press $\boxed{\text{2nd}}$ $\boxed{\text{F6}}$ $\boxed{\text{ENTER}}$. (F6 is the second operation associated with the $\boxed{\text{F1}}$ key.)

Rather than using the TRACE feature to move the cursor to a point of tangency, we can also enter the x-coordinate of the point directly. For example, to graph the line tangent to the curve at $x = 1$, from the Graph screen press $\boxed{\text{F5}}$ $\boxed{\text{alpha}}$ $\boxed{\text{A}}$ as before to select the Tangent feature. Then press 1 $\boxed{\text{ENTER}}$ to enter the x-coordinate, 1, and draw the tangent line.

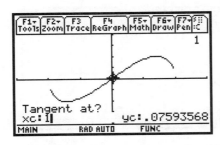 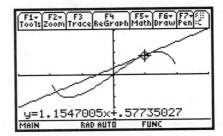

The graph of the tangent line can be cleared as described above.

If the arrow keys have been used to position the cursor at a point when using the dy/dx operation from the CALC menu, the tangent line at that point can be graphed from the GRAPH screen immediately after the value of dy/dx is displayed by pressing $\boxed{\text{F5}}$ $\boxed{\text{alpha}}$ $\boxed{\text{A}}$ $\boxed{\text{ENTER}}$.

ENTERING THE SUM OF TWO FUNCTIONS

Section 2.5, page 142 The Technology Connection on this page involves finding the derivative of a function $y_3 = y_1 + y_2$, where $y_1 = x(100 - x)$ and $y_2 = x\sqrt{100 - x^2}$. To enter y_3, first press $\boxed{\diamond}$ $\boxed{\text{Y} =}$ to go to the equation-editor screen. Clear any entries present and turn off the plots. Then enter y_1 and y_2. Note that we must include a multiplication symbol when entering y_1: $y_1 = x \times (100 - x)$. We will enter y_3 as $y_1(x) + y_2(x)$. Press $\boxed{\text{Y}}$ $\boxed{1}$ $\boxed{(}$ $\boxed{\text{X}}$ $\boxed{)}$ $\boxed{+}$ $\boxed{\text{Y}}$ $\boxed{2}$ $\boxed{(}$ $\boxed{\text{X}}$ $\boxed{)}$ $\boxed{\text{ENTER}}$.

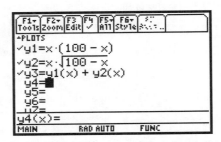

DESELECTING FUNCTIONS; GRAPH STYLES

Section 2.7, page 160 The Technology Connection on this page involves deselecting a function and also using different graph styles for two functions. First enter $y_1 = \dfrac{x^2 - 3x}{x - 1}$, $y_2 = \dfrac{x^2 - 2x + 3}{(x - 1)^2}$, and $y_3 = \text{nDeriv}(y1(x), x, x)$. We can enter "nDeriv(" by selecting this item from the CATALOG, by selecting it from the MATH menu, or by typing it directly. In each case we begin by positioning the cursor beside "y3 =." To use the CATALOG press $\boxed{\text{CATALOG}}$ $\boxed{\text{N}}$. These keystrokes display the CATALOG and move the selection cursor to the first item that begins with N. Use $\boxed{\triangledown}$ to move the cursor beside "nDeriv(" and press $\boxed{\text{ENTER}}$ to copy it to the entry line of he equation-editor screen.

To select "nDeriv(" from the MATH menu go to the home screen and first press $\boxed{\text{2nd}}$ $\boxed{\text{MATH}}$. (MATH is the second operation associated with the 5 numeric key.) Then select item A, Calculus, by pressing $\boxed{\text{alpha}}$ $\boxed{\text{A}}$ or by using the $\boxed{\triangledown}$ key to scroll down to item A and pressing $\boxed{\text{ENTER}}$. Now select item A, "nDeriv(," from the Calculus submenu by pressing $\boxed{\text{alpha}}$ $\boxed{\text{A}}$ or by scrolling down to item A and pressing $\boxed{\text{ENTER}}$.

To type "nDeriv(" directly, press $\boxed{\text{alpha}}$ $\boxed{\text{N}}$ $\boxed{\uparrow}$ $\boxed{\text{D}}$ $\boxed{\text{2nd}}$ $\boxed{\text{a-lock}}$ $\boxed{\text{E}}$ $\boxed{\text{R}}$ $\boxed{\text{I}}$ $\boxed{\text{V}}$ $\boxed{\text{alpha}}$ $\boxed{(}$.

Since we want to see only the graphs of y_2 and y_3, we will deselect y_1. To do this, use $\boxed{\triangle}$ to highlight the expression beside y1 = and press $\boxed{\text{F4}}$. Note that there is no longer a check mark to the left of y1. This indicates that y_1 has been deselected and, thus, its graph will not appear with the graphs of y_2 and y_3, the functions which continue to be selected.

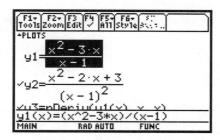 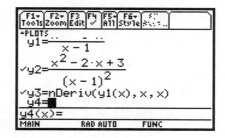

With y_1 deselected, we can graph y_2 and y_3 using different graph styles to determine whether the graphs coincide. We will keep the solid line graph style for the graph of y_2 and select the path style for the graph of y_3. After the graph of y_2 is drawn, a circular cursor will trace the leading edge of the graph of y_3 and draw its path. (This assumes that the grapher is set in Sequential mode as shown above.)

To select the path style for y_3 from the Style menu, highlight the expression for y_3 and press $\boxed{\text{2nd}}$ $\boxed{\text{F6}}$ 6 or press $\boxed{\text{2nd}}$ $\boxed{\text{F6}}$, highlight item 6, Path, and then press $\boxed{\text{ENTER}}$.

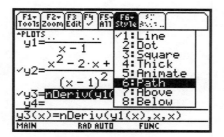

To see the graphs of y_2 and y_3 in the standard window, press $\boxed{\text{F2}}$.

To reselect y_1, highlight the expression again and press $\boxed{\text{F4}}$. There is now a check mark beside y1 again, indicating that the function is selected.

Chapter 3
Applications of Differentiation

THE MAXIMUM AND MINIMUM FEATURES

Section 3.1, page 195 Use the Maximum and Minimum features of a grapher to approximate the relative extrema of $f(x) = -0.4x^3 + 6.2x^2 - 11.3x - 54.8$.

First graph $y_1 = -0.4x^3 + 6.2x^2 - 11.3x - 54.8$ in a window that displays the relative extrema of the function. Trial and error reveals that one good choice is $[-10, 20, -100, 150]$, xscl = 5, yscl = 50. Observe that a relative maximum occurs near $x = 10$ and a relative minimum occurs near $x = 1$.

To find the relative maximum, first press $\boxed{\text{F5}}$ 4 or $\boxed{\text{F5}}$ $\boxed{\triangledown}$ $\boxed{\triangledown}$ $\boxed{\triangledown}$ $\boxed{\text{ENTER}}$ to select the Maximum feature from the Math menu on the Graph screen. We are prompted to select a lower bound for the relative maximum. This means that we must choose an x-value that is to the left of the x-value of the point where the relative maximum occurs. This can be done by using the left- and right-arrow keys to move the cursor to a point to the left of the relative maximum or by keying in an appropriate value.

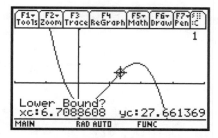

Once this is done, press $\boxed{\text{ENTER}}$. Now we are prompted to select an upper bound. We move the cursor to a point to the right of the relative maximum or we key in an appropriate value.

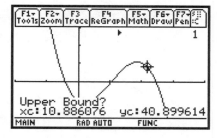

Press $\boxed{\text{ENTER}}$ again. We see that a relative maximum function value of approximately 54.61 occurs when $x \approx 9.32$.

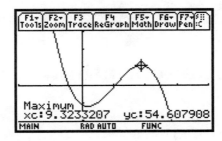

To find the relative minimum, select the Minimum feature from the Math menu by pressing $\boxed{\text{F5}}$ 3 or $\boxed{\text{F5}}$ $\boxed{\triangledown}$ $\boxed{\triangledown}$ $\boxed{\text{ENTER}}$. Select lower and upper bounds for the relative minimum as described above. We see that a relative minimum function value of approximately -60.30 occurs when $x \approx 1.01$.

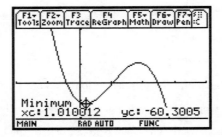

THE fMax AND fMin FEATURES

The fMax and fMin features can be used to calculate the x-values at which relative maximum and minimum values of a function occur over a specified closed interval.

Section 3.1, page 195 Use the fMax and fMin features of a grapher to approximate the relative extrema of $f(x) = -0.4x^3 + 6.2x^2 - 11.3x - 54.8$.

First enter the function as y_1 and graph it as described above. Observe that a relative maximum occurs for an x-value that is greater than 5. Now press $\boxed{\text{HOME}}$ or $\boxed{\text{2nd}}$ $\boxed{\text{QUIT}}$ to go to the home screen. Highlight the entry line and then press $\boxed{\text{F3}}$ 7 to select the fMax feature from the Calc menu and copy it to the entry line. Now we will enter the name of the function and the variable, and we will also tell the grapher to search for a relative maximum value for $x > 5$. Press $\boxed{\text{Y}}$ 1 $\boxed{(}$ $\boxed{\text{X}}$ $\boxed{)}$, $\boxed{\text{X}}$ $\boxed{)}$ $\boxed{|}$ $\boxed{\text{X}}$ $\boxed{\text{2nd}}$ $\boxed{>}$ 5. (The $>$ symbol is the second operation associated with the $\boxed{.}$ key.) Press $\boxed{\text{ENTER}}$ to find that the relative maximum occurs when $x \approx 9.32332130455$. To find the relative maximum value of the function, we evaluate the function for this value of x. Again we see that a relative maximum function value of approximately 54.61 occurs when $x \approx 9.32$.

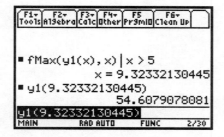

From the graph we also observe that a relative minimum occurs for an x-value less than 5. To find the relative minimum in this interval, first press $\boxed{\text{F3}}$ 6 to select the fMin feature from the Calc menu. Then enter the name of the function and the variable, and also tell the grapher to search for a relative minimum value for $x < 5$. Press $\boxed{\text{Y}}$ 1 $\boxed{(}$ $\boxed{\text{X}}$ $\boxed{)}$, $\boxed{\text{X}}$ $\boxed{)}$ $\boxed{|}$ $\boxed{\text{X}}$ $\boxed{\text{2nd}}$ $\boxed{<}$ 5. (The < symbol is the second operation associated with the 0 numeric key.) We see that a relative minimum function value occurs when $x \approx 1.01001202889$. To find the relative minimum value of the function we evaluate the function for this value of x. A relative minimum function value of approximately -60.30 occurs when $x \approx 1.01$.

Chapter 4
Exponential and Logarithmic Functions

SOLVING EXPONENTIAL EQUATIONS

Section 4.2, page 307 Solve $e^t = 40$ using a grapher.

The Intersection feature is discussed on page 127 of this manual, and the Zero feature is discussed on page 128.

EXPONENTIAL REGRESSION

Section 4.3, page 323 To find an exponential equation that models a set of data, enter the data in lists as described on page 131 of this manual. Then select ExpReg (exponential regression) as the Calculation Type, enter the lists to be used for x and y, and find the regression equation by pressing ENTER .

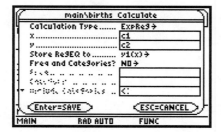

The grapher will return the values of a and b for an equation of the form $y = a \cdot b^x$. The function found can be evaluated for various values of x as described on pages 123 and 124 of this manual.

Chapter 5
Integration

THE nInt FEATURE

Definite integrals can be evaluated using the nInt feature from the Calc menu.

Section 5.2, page 384 Evaluate $\int_{-1}^{2}(x^3 - 3x + 1)dx$ using the nInt feature of a grapher.

First select the nInt feature. This is item 8 on the Calc menu. Then enter the function, the variable, and the lower and upper limits of integration. Press $\boxed{\text{F3}}$ $\boxed{\text{alpha}}$ $\boxed{\text{B}}$ $\boxed{\text{X}}$ $\boxed{\wedge}$ $\boxed{3}$ $\boxed{-}$ $\boxed{3}$ $\boxed{\text{X}}$ $\boxed{+}$ $\boxed{1}$ $\boxed{,}$ $\boxed{\text{X}}$ $\boxed{,}$ $\boxed{(-)}$ $\boxed{1}$ $\boxed{,}$ $\boxed{2}$ $\boxed{)}$ $\boxed{\text{ENTER}}$. We find that $\int_{-1}^{2}(x^3 - 3x + 1)dx = 2.25$.

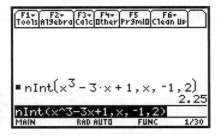

If the function has been entered on the Y = screen, as y_1 for instance, we can evaluate it by entering nInt(y1(x), x, −1, 2) on the home screen. Press $\boxed{\text{F3}}$ $\boxed{\text{alpha}}$ $\boxed{\text{B}}$ $\boxed{\text{Y}}$ $\boxed{1}$ $\boxed{(}$ $\boxed{\text{X}}$ $\boxed{)}$ $\boxed{,}$ $\boxed{\text{X}}$ $\boxed{,}$ $\boxed{(-)}$ $\boxed{1}$ $\boxed{,}$ $\boxed{2}$ $\boxed{)}$ $\boxed{\text{ENTER}}$.

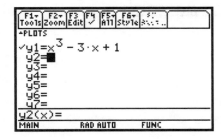 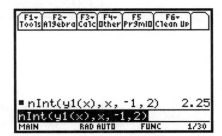

THE $\int f(x)dx$ FEATURE

Definite integrals can also be evaluated using the $\int f(x)dx$ feature from the Math menu on the Graph screen.

Section 5.2, page 384 Evaluate $\int_{-1}^{2}(x^3 - 3x + 1)dx$ using the $\int f(x)dx$ feature of a grapher.

First graph $y_1 = x^3 - 3x + 1$ in a window that contains the interval $[-1, 2]$. We will use $[-3, 3, -6, 6]$. Then select the $\int f(x)dx$ feature from the Math menu by pressing $\boxed{\text{F5}}$ 7. We are prompted to enter the lower limit of integration. Press $\boxed{(-)}$ 1.

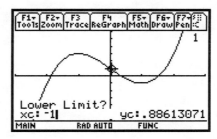

Then press ENTER. Now enter the upper limit of integration by pressing 2.

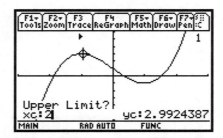

Press ENTER again. The grapher shades the area above and below the curve on $[-1, 2]$ and returns the value of the definite integral on this interval.

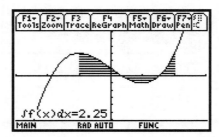

Chapter 6
Applications of Integration

STATISTICS

Section 6.8, page 475   The weights w of students in a calculus class are normally distributed with mean 150 lb and standard deviation 25 lb. Find the probability that a student's weight is from 160 lb to 180 lb.

The TI-89 does not have a built-in capability to graph a probability density function. However, the Statistics with List Editor package found under "Free Downloads" on the Texas Instruments web site, www.ti.com, provides a means for performing calculations involving the standard normal distribution. This package can be downloaded to your grapher.

Chapter 7
Functions of Several Variables

PARTIAL DERIVATIVES

Section 7.2, page 505 Given the function $f(x, y) = 3x^3y + 2xy$, use a grapher that finds derivatives of functions of one variable to find $f_x(-4, 1)$ and $f_y(2, 6)$.

To find $f_x(-4, 1)$, first find $f(x, 1)$:
$$f(x, y) = 3x^3y + 2xy$$
$$f(x, 1) = 3x^3 \cdot 1 + 2x \cdot 1$$
$$= 3x^3 + 2x$$

Now we have a function of one variable, so we use the nDeriv operation or the dy/dx operation to find the value of the derivative of this function when $x = -4$. (See pages 137 and 138 of this manual for the procedures to follow.)

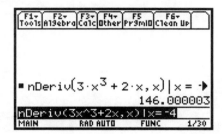   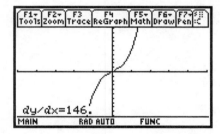

The procedures the grapher uses to calculate the derivative might not yield an exact answer. Note that the exact answer is 146, but the nDeriv operation yields 146.000003.

To find $f_y(2, 6)$, first find $f(2, y)$:
$$f(x, y) = 3x^3y + 2xy$$
$$f(2, y) = 3 \cdot 2^3y + 2 \cdot 2 \cdot y$$
$$= 24y + 4y = 28y$$

Now find the derivative of $f(y) = 28y$ when $y = 6$ using the nDeriv operation.

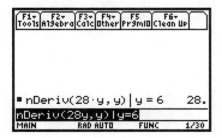

We can also replace y with x and find the derivative of $f(x) = 28x$ when $x = 6$ using the dy/dx operation.

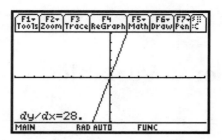

Index
TI-83 and TI-83 Plus Graphics Calculators

Index
TI-85 Graphics Calculator

Index for the TI-85 Graphics Calculator

Index for the TI-86 Graphics Calculator

Index
TI-89 Graphics Calculator

Absolute value, 129
Addition of functions, 140
Ask mode for a table, 123
Auto mode for a table, 122

Catalog, 129
Contrast, 117
Copying regression equation to
    equation-editor screen, 131
Cube root, 130
Cubic regression, 134
Curve fitting
    cubic regression, 134
    exponential regression, 147
    linear regression, 130
    quadratic regression, 134
    quartic regression, 135

Data, entering, 131
Derivative, 137, 138
    partial, 153
Deselecting a function, 140
Dot graph style, 125
dy/dx, 138

Editing entries, 120
Entering data, 131
Entry line, 117
Equations
    deselecting, 140
    graphing, 120

    solving, 127, 128, 147
Evaluating functions, 123, 124
Exponential functions, 147
Exponential regression, 147

fMax, 144
fMin, 144, 145
Function values, 123, 124
    maximum, 143, 144
    minimum, 143, 144, 145
Functions
    absolute-value, 129
    adding, 140
    deselecting, 140
    evaluating, 123, 124
    exponential, 147
    graphing, 124, 129, 130
    maximum value, 143, 144
    minimum value, 143, 144, 145
    piecewise, 124
    radical, 130
    sum of, 140
    of several variables, 153
    zeros, 128

Graph styles
    dot, 125
    path, 141
Graphing equations, 120
    points of intersection, 127
Graphing functions, 124, 129, 130

Index for the TI-89 Graphics Calculator